Gamification and Industry 4.0

Gamification and Industry 4.0 are two concepts that each in their own right have sparked attention. Gamification as a way to use game elements in non-game activities, and Industry 4.0 as a way to describe how the interconnected digital transformation of operational technologies are changing manufacturing in the 21st Century. Gamification now offers an opportunity to enhance the productivity of manufacturing and improve quality of work life in the process.

Gamification and Industry 4.0 looks at how the transformative shift of production and manufacturing opens up for gamified experiences in the work place. What the industry requires are deeper kinds of gamification, comprehending the knowledge of industrial production, computer applications, game design, learning, and motivational psychology. In order to achieve this, this book offers interdisciplinary expertise from such fields as psychology, management, software engineering, information science, game studies, and industrial production and innovation. The concept of Industry 4.0 and different types of gamifications are discussed in relation to motivation and learning. The book also examines cybersecurity of gamified systems and the potential risks any changes made to digital devices and industrial control systems may cause.

This unique book will be of value to researchers, academics and upper-level students across various fields, but in particular, management and organization studies, production and manufacturing as well as technology and innovation management.

Lars Konzack is an associate professor in Information Studies at the Department of Communication (COMM) at the University of

Copenhagen (UCPH). Konzack has an MA in Information Science and a Ph.D. in educational games from Aarhus University. He is the co-founder of the game developing academy DADIU. His fields of interest are gamification, ludology, imaginary worlds, and digital culture. Konzack has published essays on themes like educational games, role-playing games, video games, and geek culture.

Lars Konzack was the research leader of the AGAVE project, which is a project about gamification of industrial production processes. AGAVE is funded by the Innovation Fund Denmark, and it is a collaboration between the University of Copenhagen, Aalborg University, Novo Nordisk, and BioLean.

Routledge Advances in Production and Operations Management

This series sets out to present a rich and varied collection of cutting-edge research on production and operations management (POM), addressing key topics and new areas of interest in order to define and enhance research in this important field. Bringing together academic study on all aspects of planning, organizing and supervising production, manufacturing or the provision of services, subject areas will include, but are not limited to: operations research, product and process design, manufacturing strategy, scheduling, quality management, logistics and supply chain management. Highly specialised and industry-specific studies are actively encouraged.

Digitalization and Sustainable Manufacturing
Twin Transition in Norway
Edited by Sverre Gulbrandsen-Dahl, Heidi C. Dreyer, Einar L. Hinrichsen, Halvor Holtskog, Kristian Martinsen, Håkon Raabe and Gabor Sziebig

Sustainability and Smart Manufacturing
The Transformation of the Steelwork Industry
Bożena Gajdzik, Radosław Wolniak, Wieslaw Grebski, Jan Szymszal and Michalene Eva Grebski

Gamification and Industry 4.0
Gamified Smart Manufacturing
Lars Konzack

For more information about this series, please visit: www.routledge.com/The-Routledge-Philosophers/book-series/RAPOM

Gamification and Industry 4.0

Gamified Smart Manufacturing

Lars Konzack

LONDON AND NEW YORK

First published 2025
by Routledge
4 Park Square, Milton Park, Abingdon, Oxon OX14 4RN

and by Routledge
605 Third Avenue, New York, NY 10158

Routledge is an imprint of the Taylor & Francis Group, an informa business

British Library Cataloguing-in-Publication Data
A catalogue record for this book is available from the British Library

ISBN: 9781032524795 (hbk)
ISBN: 9781032524801 (pbk)
ISBN: 9781003406822 (ebk)

DOI: 10.4324/9781003406822

Typeset in Times New Roman
by Newgen Publishing UK

Contents

Figures

Acknowledgements

It is a privilege of mine to thank Innovation Fund Denmark for their support of the project AGAVE: Applied Gamification and Visualization Enhancements – for cultivating onsite operator training and improving the quality of work life. Industrial contributions have been essential in advancing gamification prototypes, and I would like to thank any and everyone who has been excitedly involved with the project.

Introduction

Attention to gamification took off when social media broke through, and it became easy to create apps for mobile phones. Now there was the possibility of tracking and providing instant feedback. It was as if all aspects of life took on a game-like character. It was a small reward point whenever someone got a like on social media. Social groups started giving each other badges. Fitness and health apps tracked progress, created small competitions, and the entire experience could be tracked on a profile where one could see their advancements. Similarly, educational applications leveraged points, badges, achievements, and leaderboards to advance learning without distinction as to whether the subject matter pertained to language or another discipline.

It was obvious that the industry could exploit gamification in their work processes, and this has happened to some extent, but only to a limited degree. At the same time, there was a revolution in the spread of artificial intelligence, robots, and network solutions for cyber-physical systems. All of this was collectively called Industry 4.0, and the new type of production was named smart manufacturing. As opportunities arose, it became possible to gamify industrial production, and right now, we are in a situation where the entire way we work is changing. The companies that can best adapt to the new production methods and become more productive will win the race to have the most well-functioning smart factories in Industry 4.0.

Over more than two decades, my academic pursuits in game studies have included entertainment products and educational games. In the late 2010s, I got involved in a research project in which we focused on creating gamification for operators at a pectin factory. Although we devised a theoretical gamification framework, the implementation did not materialize, yet the process enhanced my resolve to carry on. In 2021, we managed to secure funding from Innovation Fund Denmark for a project on gamification called AGAVE (applied gamification and visualization enhancements) for cultivating onsite operator training and improving quality of work life).

My curiosity was about how we could make the least motivated employees in the industry as motivated as the top motivated employees. One could see how video games could motivate people for hours on end. When talking about increasing productivity, engineers focused on improving technology and processes, but often the role of employees in the work process was taken for granted. This is also more complex, partly because it is hard to measure and partly because there are no simple adjustments to be made. Here, gamification becomes relevant, as a gamification solution theoretically provides the necessary tool. But it's not enough to have motivated employees. Unfortunately, there is not always a direct correlation between motivated employees and productivity. Hence, when the emphasis is placed on productivity, gamification is essentially concerned with imposing discipline among employees to ensure their tasks are conducted efficiently, which gamification can facilitate. A third curiosity is how we train using courses, reading heavy job instructions and standard operating procedures, and on-the-job training. Here, gamification grants the opportunity to conceptualize the framework of employee training in a completely innovative way. By using serious games and game-like initiatives, it is possible to create more effective employee training.

This leads to the research question: How is it possible to develop gamification and serious games that can promote motivation, productivity, and learning in the context of Industry 4.0 and smart manufacturing?

To answer this question, it is first necessary to establish what gamification and serious games are and what these gamified concepts can do. Next, it is imperative to find out what Industry 4.0 and smart manufacturing are and how these concepts function in industrial production. Then, the concepts of Industry 4.0 and smart manufacturing, as well as gamification and serious games, are aligned to categorize different types of gamification and serious games for Industry 4.0 and smart manufacturing, determining each of their potentials. Finally, this leads to a discussion and recommendations on how to develop serious games, especially gamification for Industry 4.0 and smart manufacturing.

1 Gamification

What is gamification?

Gamification is the procedure of integrating game design elements into non-game contexts by using game mechanics and experience design so that these elements increase engagement and motivation (Deterding, et al., 2011; Burke, 2014). The underlying principle of gamification posits that the incorporation of game-like characteristics into an activity or task, thereby converting it into a gamified iteration, promotes a higher inclination among individuals to engage in it and consequently encourages heightened motivation to persist until its completion. The goal of gamification is to increase the involvement, drive, and active participation in various non-game settings by applying game design techniques. The intention lies in amplifying participation, efficacy, and accomplishment as regards both individual and collective objectives (Burke, 2014).

The term 'gamify' has been attributed to Roy Trubshaw, who worked on the online game multi-user dungeons (MUD) in the late 1970s. He is recognized for introducing the concept in the context of transforming non-game elements into games. Conversely, the term 'gamify' within the gamification domain refers to employing game design techniques in circumstances unrelated to games, without fundamentally converting the non-gaming entity into an actual game (Bartle, 2016).

Computer programmer and game developer Nick Pelling coined the term gamification as a deliberately ugly word in 2002 when designing game-like interfaces for mobile phones, ATMs, and vending machines (Mazur-Stommen & Farley, 2016). Nevertheless, the word gained traction in 2010 during the 'DICE Summit' (Design Innovate Communicate Entertain) in Las Vegas when video game designer

DOI: 10.4324/9781003406822-1

Jesse Scheel warned about an apocalyptic future in which games and gaming would permeate all human life and each action would be tracked, recorded and rewarded (Risso & Paesano, 2021).

Diverse methodologies in gamification may emerge as a response to specific contexts and objectives at hand. A selection of foundational game elements within gamification might encompass:

- Setting clear goals and providing feedback to help users understand their progress.
- Providing rewards and incentives to motivate users to continue engaging with the task or activity.
- Creating a sense of competition and social collaboration to encourage users to compare their progress with others.
- Using game mechanics like levels, points, badges, and leaderboards to create a sense of progress.
- Offering personalized and dynamic content to make the experience unique for each user.
- Progress tracking and the use of metrics and feedback to monitor and communicate a player's advancement towards a goal or achievement.

The above examples show how game elements have been used (Werbach & Hunter, 2012; Burke, 2014; Chou, 2016; Chishti, 2020; Vesa, 2021). While this framework certainly offers a preliminary comprehension of game constituents, delving much deeper into games is imperative to truly apprehend the intricate tapestry woven by the elements of game design.

Game design elements

As we have already established, gamification is the process by which game design elements are incorporated into non-gaming environments. What are these game design elements is the next obvious question. It becomes vital to comprehend what a game is in order to respond to this query.

Game designer Greg Costikyan came up with the conceptual understanding that 'a game is an interactive structure of endogenous meaning that requires players to struggle toward goals.' (Costikyan, 2002, p. 21)

Costikyan posited that every game comprised five essential components: 1) interactivity, 2) a set of rules that established its structure, 3) the inclusion of competition as a form of struggle, 4) the inherent focus on goals, where the aim was to win and prevent defeat. Finally, Costikyan talked about 5) endogenous meaning, which referred to games creating their own internal logic and meaning. The game rules described through game mechanics and the game system defined how the game world functioned. Endogenous meant that something arose or came from within a given system, and regarding games, it was the meaning created from within the game system through rules and gameplay mechanics. Thus, the game's endogenous meaning may differ from the meaning in the real world (Costikyan, 2002).

Pursuing the notion of interactivity, Greg Costikyan suggests, 'What makes a thing into a game is the need to make decisions.' (Costikyan, 2002, p. 11). Interactivity is only a single part of Costikyan's game definition, and accordingly, this does not mean that every decision-making process is considered a game. However, gamification actually involves turning decision-making into a game-like experience. This means gamification alludes to altering the practice of decision-making into experiences that bear a resemblance to a game. Consequently, if a process does not have human agency and decision-making, then it is not suited for gamification.

If we consider Costikyan's elements of a game, we can understand how these game design concepts can be incorporated into contexts other than games through gamification. Let us explore each element and its implications for gamification (Costikyan, 2002):

1. Interactivity: Games possess an inherent quality of interactivity, demanding enthusiastic involvement from players. Gamification, in turn, presents a way to bring about interactivity within systems, empowering users with opportunities to engage, exercise agency, and receive responsive feedback. This engrossing experience can be artfully cultivated by employing interactive interfaces, user interfaces, or other digital platforms, facilitating active interaction between users and the gamified components.
2. Structure: Games, as complex systems of interactive engagement, rely on a framework of established rules and procedures that both delimit and stimulate the actions and possibilities of the participating players. In gamification, incorporating rules can provide

structure and guidelines for users to follow while engaging with the gamified system. These rules and procedures can be designed to align with the objectives and desired behaviors of the users, creating a framework for the user's involvement and progress.

3. Struggle: Games often involve challenges and competition, where players strive to overcome obstacles or outperform others. Gamification can leverage this element by introducing challenges, goals, and competition among users. By setting clear objectives and providing rewards or recognition for achievements, gamification can motivate and engage users in the desired activities.
4. Goals and objectives: Games have explicit goals or objectives that players aim to achieve. Similarly, in gamification, clear goals and objectives can be defined to drive user behavior and engagement. By establishing meaningful goals and providing a sense of progress or accomplishment, gamified systems can motivate users to participate and continue their journey.
5. Endogenous meaning: Games create their own internal logic and meaning through rules, game mechanics, and systems. This can be employed by aligning the gamified elements with the underlying purpose or domain, incorporating narratives, themes, or visual representations that resonate with the users.

While Greg Costikyan offers a definition of the essential elements that constitute a game, it is important to acknowledge that games can go beyond these minimum requirements. Game design is a dynamic field that constantly explores new components to enhance the gaming experience. Contemporary video games frequently include features like storytelling, visually engaging graphics, carefully designed sound, progressive character evolution, intricate in-game environments, ethical decision-making, and opportunities for social engagement, among various other aspects (Fullerton, 2018; Moore, 2011).

Engaging in gameplay as a gamer can involve various experiences that bring satisfaction. These experiences can range from exploring virtual worlds to developing strategic thinking and working together to solve problems in digital environments. Experiences can be grouped into eight categories: sensation, fantasy, narrative, challenge, fellowship, discovery, expression, and submission (Costikyan, 2002; Hunicke, et al., 2004). Sensation involves the aesthetic experience of hearing music, seeing beautiful visuals, or using empowering control schemes. Fantasy involves immersing oneself in a grand fictional world and imagining oneself as part of it. Narrative is the aspiration that comes from the dramatic unfolding of a sequence of events through a game's story. Challenge is the pleasure of accomplishing tasks or solving puzzles. Fellowship involves the joy of feeling a sense of community or friendship while playing. Discovery is the delight of exploring the game world or finding a secret feature or strategy. Expression involves the satisfaction of creating or expressing oneself, such as character creation or community-level building. Finally, submission is the basic transaction that occurs when we play a game, as we submit to its structure and rules.

The next question is, of course, which game elements are important for gamification. This may, of course, depend on the actual goal of the gamification application and may, therefore, not have a finite answer. Nevertheless, a study with 19 experts (including researchers, teachers, game developers, a designer, and an artificial Intelligence engineer) has been conducted to come up with a possible answer (Toda, et al., 2019). In this study objectives, level, and progression were identified as the most important game elements for education. However, in hindsight, it was suggested to include two additional game elements, narrative and storytelling, which were considered important due to their connection to human behavior and the need for storytelling. Another answer to this question comes from a study of interaction design, pointing towards nine typical elements: points, badges, leaderboards, progress bars, performance graphs, quests, meaningful stories, avatars, and profile development (Sailer, et al., 2013). This tells us that these elements are often used in gamification but, unfortunately, not how effective they are in fostering motivation.

Serious games versus gamification

Gamification and serious games are two terms that are frequently used interchangeably; however, they refer to two separate ideas. Games with a primary focus on something other than pure enjoyment, such as education, training, or simulation, are referred to as serious games (Djaouti, et al., 2011). Gamification, as already mentioned, refers to the use of game elements in non-game contexts to motivate and engage users. This means that they both have a grounding in game design, but the method is different. The serious games approach is to change an entertaining game into a serious practice such as education, training, and simulation while the gamification approach is to change a serious practice by adding game elements.

One may contend that the more serious a serious game gets, the less like a game it is, and the more game features are present in a non-game environment, the more that context resembles a game. This means, at some point it becomes hard to explain and clarify the distinction between serious games and gamification. It depends on what the starting point was rather than the actual outcome. In this sense, gamification is in the middle of a transition between a serious activity and an entertaining game.

Both gamification and serious games have distinct advantages. While gamification can encourage and engage users in situations other than games, serious games offer a fully immersive and engaging experience for training and educational purposes. The distinction between these two ideas will probably become hazier as technology develops, and more hybrid strategies might even be implemented. Any serious game or gamification project's ability to engage and inspire users toward a particular goal or objective ultimately determines whether it will be a success.

A short pre-history of gamification

While Nick Pelling coined the term gamification in 2002, the idea of using games and game-like incentives in non-game contexts is much older (Goethe, 2019). Though it must be added that even though these incentives had some of the game-like properties, they were often superficial.

In the 19th century, Prussia invented Kriegspiel (war-game) for the military. It is commonly ascribed to Georg von Reisswitz, a Prussian army officer, who in 1824 created the wargame *Anleitung zur Darstellung militairischer Manöver mit dem Apparat des Kriegs-Spieles* (Instructions for depicting military maneuvers with the apparatus of the war game) based on military tactics, maps and probability (Wintjes, 2016). Originally, Kriegspiel was introduced to Prussian officers as a didactic tool. Its principal purpose was to aid officers in acquiring the art of strategic thinking and decision-making in a simulated combat setting. The game swiftly gained traction among officers and was subsequently adopted by armies around the globe (Hilgers, 2012). Kriegspiel's development by the Prussians marks a turning point in the history of military instruction and strategy. It is impossible to overstate the virtue of the game as a teaching tool since it allowed officers to practice strategy and judgment in a simulated battlefield setting. Prussian wargames were designed as serious training tools for military officers, rather than as gamification. Despite the military being a non-game context, the wargames were more than just a compilation of game elements; they constituted a complete game. Officers were trained in strategy, tactics, and quick decision-making through the use of these games, which simulated real-world warfare events. The aim of the endeavor was to foster the officers' critical thinking and problem-solving abilities while simultaneously affording them a pleasurable pursuit.

Early gamification came about when the pioneering concept of S&H Green Stamps appeared in 1896, thanks to the ingenuity of Thomas Sperry and Shelley Byron Hutchinson (Pollack, 1988). The privilege of collecting these stamps and exchanging them for a variety of goods listed in the redemption options was available to the clients. Customers felt a feeling of accomplishment and excitement as they collected and traded their stamps, which helped the program promote repeat business and brand loyalty.

The year 1908 was the birth year of the Boy Scout movement. Led by British Army officer Robert Baden-Powell, it ushered in a new approach to celebrating the achievements of its members in the form of a military-inspired badge system. The Scouts were awarded badges symbolizing their skills and dexterity in various areas, their commitment to the organization's principles, and their active participation in special events. Thus, the Boy Scouts brought forth an innovative way to celebrate individual growth and progress through the symbolic power of insignia (Wilkinson, 1969; Jeal, 1989). This idea of a badge system would later prove influential in modern gamification.

According to Mark J. Nelson, the Soviet Union (1922–1991) promised a way to motivate workers without capitalist monetary incentives (Nelson, 2012). This led to Vladimir Lenin's idea of socialist competition, in which *subbotniki* (groups of workers) would compete against each other to earn points and receive awards for reaching milestones, as well as medals, e.g., the Order of the Red Banner of Labour. Conversely, Joseph Stalin began the industrialization of the USSR with more persuasion and coercion to achieve his goals. These methods were based on material rewards with the appeal of performance wages and were much closer to bourgeois competition among workers. Nevertheless, Stalin labeled the concept of Socialist Emulation (Deutscher, 1951–1952).

During the Cold War, the United States also engaged in the development of gamification. The Disney Corporation launched the 'Magic Kingdom Club' idea in the United States in 1957 with the goal of fostering steadfast patronage among its customers. This was a calculated move to keep customers coming back and cultivate a devoted fan base for Disney's burgeoning theme park empire. Members received a variety of perks and rewards, such as discounted admission to Disney theme parks, special goods, and early access to new attractions (Smith, 1998; Strodder, 2012). The AAdvantage loyalty program was introduced by American Airlines in 1981. This program compensated members for their trip and was initially only available to members who received invitations. A first-class voucher was made available to members for travel on any American Airlines route. With incentives for frequent use, the program aimed to promote consumer loyalty. Within a few years, every significant airline in the United States imitated the practice as it gained popularity (Mak, 2003).

This tells the story of how non-game practices were gamified at a time before social media and before computers and the Internet had become commonplace. The emphasis was on providing incentives to influence people's behavior in the preferred direction. Teachers have traditionally used marks and grades to motivate their pupils through positive reinforcement, and many businesses and organizations have fashioned their own initiatives with a variety of rewards to encourage customer loyalty or other types of behavioral adjustments (Bloom, 1977; Ulrich, 1989).

Due to the domestication of computers into private homes, the expansion in internet usage, and the normalization of smartphones into everyday routines, if not common habits, digital gamification has developed, especially with social networking, fitness and health, and educational apps (Zicherman & Cunningham, 2011). The history of digital gamification is built on the emergence of commercial video games in the late 20th century and earlier incentive programs.

Digital gamification

The first commercial video games appeared in the early 1970s, and at it was at that time in 1973, Charles A. Coonradt published The Game of Work, in which things like measurement and scorekeeping, goals and objectives, rewards and recognition should make the basis for creating continuous improvements and a winning culture at the workplace (Coonradt, 2012). Although this concept was not digital, it laid the ground for thinking of work life as a game.

According to the prevailing interpretation of gamification, the incorporation of game-like features into non-gaming contexts requires digital technology. Like Coonradt before him, Thomas Malone acknowledged this towering trend. Back in the late twentieth century, computer systems were hard to understand and often had confusing, misleading, and in many cases counter-intuitive interfaces (Grudin, 1990; Visser, 2006). Watching how video games of the 1970s and 1980s succeeded by not only creating understandable interfaces and getting a young audience wanting to use pocket money to play these digital games, Thomas W. Malone wanted to know what mechanisms make the learning process enjoyable. Additionally, he probed the potential for extracting valuable knowledge from video game interfaces (Malone, 1980; Malone, 1982).

In the insightful analysis put forth by Thomas W. Malone, we face the discernment of three pivotal elements that generate both gaming experiences and user interfaces as captivating and enjoyable engagements. These three factors are finding the optimal level of challenge, incorporating elements of fantasy and make-believe, and sparking curiosity (Malone, 1980; Malone, 1982). In the domain of crafting digital playful experiences and interactive systems, it's crucial to clearly outline meticulous goals, establish rapid and enlightening communication loops, and strike a balance between achievable challenges; furthermore, it's vital to embed adaptability into the fabric of the process, acknowledging its inherently fluid nature, especially when embarking on creating computer games and user interfaces, where defining precise objectives takes the spotlight, quick and perceptive feedback mechanisms are imperative, and maintaining an acceptable level of challenge is key to holding interest. To fully engage and fascinate consumers, it is also advantageous to develop an immersive atmosphere, encourage social interaction, and incorporate storytelling aspects (Bødker, 1991; Björk & Holopainen, 2005; Laurel, 2013; Fullerton, 2018). The notion is to simply adopt the captivating features of computer games and apply them to non-game interfaces.

During the 1990s, video games became more focused on not only better graphics and speed but also incorporating advanced game mechanics and storytelling (Donovan, 2010). Alongside this development, the Internet became more prevalent in society, and in 1999, Stephen W. Draper examined how the increasing accessibility of the Internet and digital resources could transform learning environments, making education more interactive and personalized with inspiration coming from video games and the motivational psychology of games (Draper, 1999).

In 2005, Rajat Paharia founded Bunchball, marking the inception of the first modern gamification platform (Hugos, 2012). Although the term gamification was not initially used, Bunchball introduced novel game mechanics such as points, badges, and leaderboards to amplify user engagement on online platforms. Bunchball, as an entrepreneur in gamification evolution, catalyzed the pervasive assimilation of gamification technology. As a result of Bunchball's accomplishments, other digital platforms emerged and this gave rise to a larger trend of gamifying marketing, employee involvement, and educational apps.

Simultaneously, major social media platforms discerned the potential of infusing game-like features into their interfaces to foster heightened user interaction (Martínez-López, et al., 2022). The strategic deployment of likes, shares, and comments metamorphosed these platforms into dynamic arenas where users not only connected but also competed for digital accolades. The intentional blend of social connectivity and gamification strategically played a vital role in sustaining and boosting user engagement across a range of online communities.

A revolutionary fusion of technology and physical activity occurred through the pioneering gamification approach by Nike+ (Burke, 2014). By seamlessly blending mobile applications with wearable devices, Nike+ not only monitored physical endeavors but also leveraged gamification to incentivize and reward users for their commitment to an active lifestyle.

The pinnacle of gamification's potential was perhaps epitomized by the meteoric rise of Pokémon GO (Hulsey, 2019). Rolled out to exceptional global acclaim, this augmented reality (AR) sensation not only fascinated millions but also emphasized the unrealized potential of location-based gamification. Pokémon GO, through cleverly blending the virtual world with actual locations, not only broke records but also demonstrated the significant influence that gamification, augmented reality, and location-based features can have when combined.

The evolution of digital gamification can be traced from the early days of commercial video games in the 1970s to the contemporary integration of game-like elements into various non-gaming contexts. Charles A. Coonradt's 1973 publication, 'The Game of Work,' laid the groundwork for viewing work as a game, emphasizing measurement, scorekeeping, goals, and rewards. Thomas W. Malone's analysis in the late twentieth century identified three crucial elements for engaging gaming experiences and user interfaces: optimal challenge level, elements of fantasy, and sparking curiosity. Video games in the 1990s evolved in terms of graphics, speed, and storytelling, a development that coincided with the widespread adoption of the Internet. Foursquare's 2009 introduction of location-based check-ins added gamification to social networking, and Badgeville in 2010 further solidified the gamification trend. Social media platforms incorporated game-like features, and Nike+ blended technology and physical activity. The sudden craze in 2016, Pokémon GO, underscored the potential that emerges when gamification, augmented reality, and location-based gaming are seamlessly combined.

Motivational psychology

Exploring motivational psychology involves identifying the triggers that prompt or trigger specific behaviors. Motivation, as expounded in the field of psychology, comprises four integral elements: 1) initiation, 2) direction, 3) magnitude, and 4) intensity and persistence in relation to a given behavior (Brown, 2007). Motivation concerns the inclination and preparedness to partake in actions, including both long-term and short-term goals.

One fundamental theory in motivational psychology is the Self-determination theory. It posits that motivation stems from three underlying psychological requirements: 1) autonomy, 2) competence, and 3) relatedness (Cerasoli, et al., 2016; Bauer & McAdams, 2000). Autonomy refers to one's need for control over their life and actions; competence pertains to feeling capable and effective, while relatedness involves the need for social bonds and relationships.

Within the field of motivational psychology, a central theory comes into focus – the theory of goal setting, which suggests that people are driven by specific and challenging objectives (Locke, 1976; Lunenburg, 2011). When individuals choose to pursue clearly defined goals that can be easily measured, akin to following a GPS system, and are as difficult as attempting to train a cat to retrieve an object, and they receive ample feedback and assistance, their motivation tends to increase significantly, leading to improved performance.

Determining precise motivators is complicated by the influence of personality, emotions, beliefs, and self-esteem on motivation, allowing for multiple interpretations (Ford & Smith, 2020). Also, motivation can be either positive or negative, leading to approach or avoidance behavior. Within the expansive field of motivational psychology, there is an ongoing process of evolution. The comprehension of individuals' motivations lies at the core of supporting them in their goal-setting and attainment while enhancing their general well-being.

Gamification is based on the tenets of motivational psychology as it incorporates game design components such as points, badges, and leaderboards to heighten involvement and drive in non-game milieus. A substantial facet of gamification that is founded on motivational psychology is the utilization of incentives and rewards (Lecture, 2002). The power of rewards as motivators is well-established, and gamification employs rewards such as points, badges, and leaderboards to provide concrete feedback on progress and to motivate users to continue their engagement in the task or activity.

Another aspect of gamification that is grounded in motivational psychology is the employment of lucid objectives and progress monitoring. Clearly defining goals and delivering feedback on progress is instrumental in gamification, aiding users in comprehending their advancements, receiving encouragement, and, optimistically, attaining a sense of fulfillment and triumph (Marcusson, 2020). This dovetails with the Self-determination theory, which posits that autonomy and competence are elemental human motivational requisites. Gamification likewise fosters a feeling of rivalry and communal cooperation by permitting users to juxtapose their progress with others. This aligns with the Self-determination theory since relatedness is the necessity for social ties and associations with others. It is also consonant with the goal-setting theory that proffers that people are galvanized by explicit and formidable goals, and juxtaposing oneself with others can generate a challenge. Furthermore, gamification proffers customized and variable content to fashion a unique experience for each user.

Extrinsic motivation and intrinsic motivation are opposing categories used to explain why individuals choose to partake in specific behaviors (Hennesey, et al., 2015). Extrinsic motivation proceeds from exterior stimuli such as rewards or penalties. Behavior impelled by extrinsic motivation is steered by the anticipation of a tangible reward or punishment, such as monetary compensation, grades, or approval. For instance, a scholar may scrutinize for an exam to achieve a respectable grade, or a laborer may toil diligently to attain a promotion. Extrinsic motivation can wield substantial influence, yet it can also prove less effective over time than intrinsic motivation (Reiss, 2012).

Contrarily, intrinsic motivation stems from the internal urge to participate in an endeavor because it is intrinsically pleasurable or fulfilling (Reiss, 2012; Hennesey, et al., 2015). Intrinsically motivated behavior is propelled by the enjoyment or gratification of the activity itself, instead of the prospect of an external reward. For example, a musician may perform an instrument because she derives pleasure from playing, or a jogger may run because she relishes the sensation of being physically active. Intrinsic motivation is deemed more sustainable and advantageous than extrinsic motivation over the long haul since it is founded on inner desires, instead of external incentives. Over an extended period, intrinsic motivation is typically perceived as the more sustainable and beneficial type of motivation.

It's important to mention that the two types of motivation are not incapable of coexisting. Individuals may possess both intrinsic and extrinsic motivation concurrently in various tasks and circumstances, and the interplay between these motivations can affect their level of involvement and achievements.

Flow is another way to understand motivation. According to Mihaly Csikszentmihalyi, flow is what happens when challenges and skills match (Csikszentmihalyi, 1991; Draper, 1999). This means that if a person is not very skilled at a task and meets challenges that require a very skilled person, this person experiences anxiety. On the other hand, if a very skilled person gets very easy challenges, then this person becomes bored. Flow theory can be used to set the right level of difficulty. Flow works particularly well as a development process if the next challenge lies just within the Zone of Proximal Development (Basawapatna, et al., 2013). The Zone of Proximal Development is the range of tasks that a learner can perform with guidance but not yet independently, and in the case of gamification, the game system guides the user (Brito, et al., 2023).

Another way to approach motivation comes from game design scholar Richard Bartle developed four player types in the context of online multi-user dungeons (MUDs): killers, achievers, socializers, and explorers (Bartle, 2006). If we view the workers in a factory through the lens of Bartle's categories, they would be classified as killers, achievers, socializers, and explorers. These player types have also been used to understand how users relate to gamification (Kocadere & Çağlar, 2018).

To recapitulate, gamification is underpinned by motivational psychology since it employs features such as incentives, progress monitoring, lucid objectives, rivalry, and personalization to stimulate involvement and motivation. In the domains of education, healthcare, employee training, and customer engagement, designers, with an understanding of motivational psychology principles, can develop game-like attributes that effectively encourage user participation in tasks or activities. In addition, the game design community has developed player types based on motivational preferences of playing games.

Shallow, deep, and balanced gamification

Two approaches exist within the domain of gamification, known as shallow gamification and deep gamification, that differ in the extent to which kinds of game design elements are incorporated into non-game contexts (Mozelius, 2021). Shallow gamification is considered an instrumentalist approach, where game mechanics are used as a means to an end, while deep gamification is likened to an idealist style of game design, seeking to create an immersive experience that incorporates game mechanics as integral components of the non-game context (Egenfeldt-Nielsen, et al., 2016).

Shallow gamification is a simplified version of gamification that uses a few basic game mechanics to motivate and engage users. This type of gamification was coined BLAP (an abbreviation of badges, leaderboards/levels, achievements, and points) by Scott Nicholson because it includes a point system, badges, leaderboards, levels, and achievements, providing users with immediate feedback and rewards (Nicholson, 2012; Nicholson, 2015).

BLAP gamification aims to encourage user behavior by providing users with small, practical prizes for taking part in a specific action or completing a task (Nicholson, 2014). An illustration of this would be a fitness software that grants users badges or points for hitting particular milestones or for completing daily workouts. The addition of the leaderboard feature could create a sense of competitiveness among users and spur them on to work even harder (Jozani, et al., 2018).

While BLAP gamification can be an effective method for motivating and encouraging users, it must be added that it also has its limitations. The rewards offered through this type of gamification are often superficial and may not be enough to sustain long-term engagement and commitment. Users may, over time, lose interest once they earn all the available badges or climb to the top of the leaderboard (Nicholson, 2014; Söbke, 2019).

The fact that BLAP gamification may not be appropriate for all tasks or behaviors must also be acknowledged. For example, this type of gamification may not be the greatest choice for tasks that call for creative or analytical thinking. In certain circumstances, it might be necessary to incorporate more intricate game mechanics to adequately stimulate and inspire users.

Deep gamification, on the other hand, is an avant-garde approach, transcending mere superficial game components by integrating game design creeds into the core of an activity or task. According to Pedro A. Santos, deep gamification 'can be defined as introducing game elements that change the core processes of the activity' (Santos, 2015, p. 2) and Henrik Söbke points out that deep gamification 'is considered to promote longer-lasting engagement than shallow gamification' (Söbke, 2019, p. 375).

Deep gamification prioritizes the development of a noteworthy and absorbing user experience. This connotes that the game aspects are intimately interrelated to the task or activity, and are fabricated to enrich the user's comprehension, enthusiasm, and motivation towards the task (Gurjanow, et al., 2019; Sánchez, et al., 2021). The notion of deep gamification centers on the conception of a game so immersive that users voluntarily assimilate it into their daily routines, making positive behavioral changes that persist over the long haul (Söbke, 2019; Mozelius, 2021; Vesa, 2021). This form of gamification transcends simplistic reward and punishment schemes, instead pivoting towards engendering a sense of mastery, autonomy, and relatedness that stimulates users to persistently engage with the game-like experience (Deterding, 2014; Mozelius, 2021).

The desire for deep gamification in the domain of game design can be compared to a pursuit of the Holy Grail. It serves as a representation of the highest ambition in crafting a completely immersive gaming encounter that seamlessly merges with the user's professional and daily schedules. Whilst conventional gamification is reliant on superficial game elements, such as points, badges, and leaderboards, to incentivize behavioral change, deep gamification strives to tap into the fundamental psychological means that infuse games with their captivating and addictive qualities (Santos, 2015; Söbke & Londong, 2019; Mozelius, 2021).

Developing a high level of competence in deep gamification is a convoluted and intimidating task. Crafting a game that blends seamlessly into users' everyday lives and work obligations without crossing over into intrusiveness or being unduly taxing poses several hurdles. Striking the correct balance between user engagement and privacy concerns is likewise a considerable challenge (Mavroeidi, et al., 2019). Gamification should aim to enhance user experiences without being overly invasive and with a focus on maintaining long-term sustainability. Users may also lose interest if the game becomes monotonous or lacks ongoing challenges and rewards (Conway, 2014). Seamlessly integrating gamification into work obligations without disrupting productivity can likewise be tricky (Diefenbach & Müssig, 2019).

While shallow and deep gamification may seem like two opposing methods, they may actually be merged, creating a more comprehensive and effective gamification strategy (Mozelius, 2021). By melding shallow gamification's rudimentary game mechanics with the immersive features of deep gamification, designers can cultivate experiences that promote user reengagement. Rather than producing simplistic shallow gamification or seeking the Holy Grail of deep gamification, the pursuit of a balanced approach with pragmatic and reasonable results based on responsible gamification strategies may be more beneficial. In this sense, gamification is a continuum between shallow and deep gamification (Lieberoth, et al., 2015).

Balanced gamification is a methodologically dynamic approach that strategically maximizes the advantages derived from both shallow and deep gamification. It ingeniously merges the motivating components inherent to shallow gamification with the all-encompassing and intentional facets characteristic of deep gamification. The final product is a gaming situation designed to retain player engagement and motivation through the provision of progress, achievement, purpose, and meaning.

Extrinsic and intrinsic motivation in shallow and deep gamification

Shallow gamification corresponds to extrinsic motivation because it relies on the use of surface-level game elements, such as points and badges, to increase engagement and motivation without creating a truly immersive and meaningful experience (Goethe, 2019).

Shallow gamification is an approach that uses game design elements, such as points and badges, in a superficial way without much thought or planning, to simply add a game-like layer to a task or activity. The goal of shallow gamification is typically to increase engagement and motivation, but it does not necessarily enhance the task or activity in a meaningful way. The focus is more on the external rewards, the points, and badges, rather than the engagement in the task itself. In accordance with the notion of extrinsic motivation, which is rooted in external elements like rewards or penalties. Behavior, driven by extrinsic motivation, hinges on the anticipation of concrete rewards or punishments, like monetary gains, grades, or commendations. In the case of shallow gamification, the rewards are the points and badges, they are the external factors that motivate the users to engage in the task.

It's important to highlight that although surface-level gamification can achieve some success in enhancing engagement and motivation, it might not match the effectiveness of profound gamification when it comes to crafting a genuinely immersive and significant experience. This is because shallow gamification does not tap into the user's intrinsic motivation and may not lead to a more sustainable engagement and a more genuine interest in the task over the long term.

Shallow gamification corresponds to extrinsic motivation by using surface-level game elements, such as points and badges, to increase engagement and motivation without creating a truly immersive and meaningful experience shallow gamification relies on external incentives and doesn't tap into the user's inner drive.

Deep gamification, on the other hand, is associated with intrinsic motivation because it generates an experience that is in line with the user's interests and goals, which ought to be engaging for the individual (Goethe, 2019). This means deep gamification predominantly revolves around creating a meaningful and engaging user experience. This involves closely integrating game elements with the task or activity, with the intent of enriching the user's comprehension, enthusiasm, and drive towards the task. This harmony corresponds with the idea of intrinsic motivation, rooted in the internal urge to engage in an activity due to its inherent enjoyment or fulfillment. Through furnishing a substantial and captivating encounter, deep gamification has the ability to access the user's intrinsic motivation, fostering a greater likelihood of engagement and commitment in task completion.

Furthermore, deep gamification allows users to have autonomy, a sense of purpose, and gives them a sense of control over their own learning and development, which is in line with the Self-determination theory. The foreseeable outcome is a more sustainable engagement and a more genuine interest in the task. Therefore, deep gamification correlates with intrinsic motivation by forging a significant and captivating encounter that harmonizes with the user's preferences and aspirations.

Counterproductive incentives

Counterproductive incentives, sometimes known as perverse incentives, are systemic inducements where the outcomes are unintended or directly undesirable consequences. These incentives can arise unintentionally when poorly thought-out consequences emerge from a particular inducement, or when some users of the system find easy ways to gain advantages without achieving the desired effect of the incentive. This phenomenon is known as 'gaming the system.'

It is important to note that gamification does not work for all situations and should be used thoughtfully and carefully for it could backfire if not implemented properly. One way gamification might fail comes from the concept of counterproductive incentives (Qayum, et al., 2014; Diefenbach & Müssig, 2019). Counterproductive incentives are incentives that have the opposite of the intended effect or that can create unintended negative consequences. These types of incentives can be in the form of rewards or punishments, but the outcome is detrimental instead of beneficial (Korine, 2022).

An example of a counterproductive incentive is the so-called cobra effect. The term was originally coined by economist Horst Siebert (Siebert, 2001). The story goes that the British government in Delhi wanted to eradicate venomous snakes and decided to offer a bounty for each dead snake. Initially, many snakes were killed, making the program seem like a success in the short term. However, people soon began breeding snakes to collect the bounty payments. When the government realized what was happening, they terminated the bounty program, leading the snake breeders to release their now-worthless snakes. As a result, there were more venomous snakes than before the program was implemented.

An instance of counterproductive incentives can be observed when a company offers rewards to its employees contingent on the quantity of sales they generate. This could potentially result in employees taking shortcuts or participating in morally questionable actions to boost their sales numbers. Consequently, the company's image and financial results may suffer negatively. Another example of counterproductive incentives is when a school rewards students for getting high test scores by giving them a prize. This may result in students engaging in dishonest behavior or disregarding other crucial facets of their education, like creativity and analytical thinking, solely to attain a high score (Prestopnik, et al., 2017; Manheim, 2023).

It is imperative to meticulously contemplate the potential repercussions of any incentive initiative to steer clear of counterproductive outcomes. It is of significance to verify that these incentives steer clear of promoting unfavorable or morally questionable conduct, as well as avoiding any adverse aftermath (Zicherman & Cunningham, 2011). Consequently, in gamification, counterproductive incentives refer to rewards or punishments that have the opposite of the intended effect or that can create unintended negative consequences.

Avoiding counterproductive consequences in gamification necessitates a thorough contemplation of the potential negative outcomes of incentive initiatives. Initiatives must be designed to align with the fundamental task or mission, being careful not to encourage detrimental or ethically questionable behavior and preventing any resulting unfavorable outcomes. Verifying that the game components integrated are in harmony with the game's goals and cater to the players' needs is indispensable, with a parallel emphasis on ensuring that the educational or engaging aspect of the encounter remains unaffected.

Learning theory

Leveraging insights from various pedagogical paradigms, the incorporation of ludic elements and principles into the assimilation of novel information, proficiencies, and conduct defines gamification in the domain of learning theory, and with this interdisciplinary method seeking to increase engagement and motivation in educational contexts (Kapp, 2013; Romero, 2020; Costello, 2020). Through the integration of game-like features such as rewards, challenges, and interactive encounters, the objective of gamification is to fine-tune the learning progression, heightening both its enjoyment and efficiency. This lively perspective on learning theory, as seen through the lens of gamification, endeavors to exploit the inherent motivational aspects of games for the development of a more immersive and captivating learning atmosphere.

Classical conditioning, a well-known learning theory proposed by Ivan Pavlov, explicates how organisms learn to respond to specific stimuli through the establishment of associations with other stimuli (Clark, 2021). E.g., a dog comprehends how to produce saliva upon hearing a bell due to its capacity to link the bell with nourishment. Dale Carnegie's influential though anecdotal book 'How to Win Friends and Influence People' from 1936 accentuated the usefulness of positive reinforcement. The fundamental principle of his approach was aligned with fostering sincere social connections thereby achieving mutual respect. Rather than commanding admiration, the leader should encourage people to do better thereby changing attitudes and winning their respect and approval (Parker, 1977; Bassford & Molberg, 2013).

Building on Pavlov's classical conditioning, B.F. Skinner, a psychologist, put forth the concept of operant conditioning. Behavior is shaped by its outcomes. Actions resulting in favorable outcomes are more likely to be repeated, whereas those resulting in unfavorable results are less probable to be performed again. B.F. Skinner introduced the concept of operant conditioning in 1953, revealing that behavior could be adjusted through positive and negative reinforcement (Skinner, 1953). Cognitive models unraveling the nuances of information processing and cognitive thought ruthlessly challenged behaviorists, blinded by their shallow emphasis on observable actions. This critical transition demarcated the shift from behaviorism to cognitive psychology in the 20th century. In the domain of cognitive psychology, an instrumental figure is Jean Piaget, whose research in the cognitive development of children is widely acknowledged, alongside Ulric Neisser, a trailblazer who inaugurated the concept of cognitive psychology in 1967 (Rieber, 1983).

Learning, as per cognitive theories, is an active endeavor where individuals utilize mental processes like attention, memory, problem-solving, and decision-making to construct their understanding of the world (Rudmann, 2017). Emphasizing the role of social and cultural influences, social learning theory proposes that individuals learn by observing and imitating others. For instance, a child learns to walk by mimicking other children. According to the constructivist learning theory, knowledge is gained through individuals actively constructing their comprehension of the world through immersive experiences and engagement with their environment (Pritchard & Woollard, 2013). Multiple factors can influence the process of learning, as it is not confined to a singular theory, however, different theories may offer different insights and perspectives on how people learn and how to facilitate learning in different settings.

Gamification ought to be based on learning theory because it uses game design elements to create an engaging and interactive learning experience (Kapp, 2013). One way that gamification is based on learning theory is using clear goals and progress tracking, and this corresponds with the goal-setting theory, which suggests that individuals are driven by distinct and demanding objectives. Explicitly outlining goals is necessary, offering users feedback to measure their progress. In doing so gamification may assist learners to comprehend any advancements and experience a sense of fulfillment and achievement. Enhanced motivation towards task completion may subsequently amplify the retention and comprehension of the subject matter.

In accordance with learning theory, gamification derives a foundation from rewards and incentives, and in the context of educational strategies, gamification deploys rewards, as well as incorporating points, badges, and leaderboards, in order to convey concrete feedback on progression, thereby encouraging the motivation of learners to engage with the assigned task or activity. The compatibility with operant conditioning theory is evident, positing that behavior is linked to the consequences it encounters. In this context, positive outcomes, exemplified by rewards, possess the potential to instigate heightened engagement and an inclination for the recurring execution of the given activity.

Gamification also incorporates elements of social learning theory, which emphasizes the role of social and cultural influences on learning (Kim, et al., 2017). Learning through observation and imitation is promoted in a gamified setting, where competition and social collaboration play key roles. The individualized and ever-changing content in gamification may be designed to offer a distinct experience for each user (Jahn, et al., 2021). The alignment with the constructivism learning theory is evident, as it posits that individuals learn by actively constructing their understanding of the world through experience and interaction with their environment (Costello, 2020).

Summary

Since the beginning of the 21st century and onwards, the path of gamification has depicted the intricate interdependence among technology, psychology, and human involvement, impacting a diverse array of domains. Coined by Nick Pelling, gamification has evolved from the foundational principles introduced by trailblazers such as Charles A. Coonradt and Thomas W. Malone to its contemporary integration of game-like elements into non-gaming contexts.

The convergence of digital gamification and motivational psychology has given rise to two contrasting yet influential methodologies: shallow gamification, relying on basic game mechanics for motivation, and deep gamification, which seeks to redefine processes through seamless integration of game design principles and fundamental psychological motivators. A third perspective, balanced gamification, strategically merges the motivating aspects of shallow gamification with the intentional facets of deep gamification, aiming to foster sustained user engagement by blending progress, achievement, purpose, and meaning.

Nevertheless, one must be cautious about potential counterproductive incentives that may undermine the positive effects of carefully implemented gamification. A delicate equilibrium between user engagement, privacy considerations, and a seamless blend into everyday life is deemed essential, and it needs to be acknowledged that embedding gamification in learning theory is quite crucial.

The congruence with various pedagogical paradigms holds the prospect of promoting increased motivation within educational environments. In the ever-evolving landscape where technology, psychology, and human engagement intersect, the careful and planned use of gamification principles appears to hold the potential for generating experiences that are transformative and engaging, be it in digital or real-world interactions.

Bibliography

Bartle, R., 2006. Hearts, Clubs, Diamonds, Spades: Players who Suit MUDs. In: *The Game Design Reader: A Rules of Play Anthology.* Cambridge, MA: MIT Press, pp. 754–787.

Bartle, R. A., 2016. MMOs from the Outside. In: *The Massively-Multiplayer Online Role-Playing Games of Psychology, Law, Government, and Real Life.* New York, NY: Apress.

Basawapatna, A. R., Repenning, A., Koh, K. H. & Nickerson, H., 2013. *The Zones of Proximal Flow: Guiding Students Through a Space of Computational Thinking Skills and Challenges.* San Diego, CA: ACM, pp. 67–74.

Bassford, T. E. & Molberg, A., 2013. Dale Carnegie's leadership principles: Examining the theoretical and empirical support. *Journal of Leadership Studies*, 6(4), pp. 25–47.

Bauer, J. J. & McAdams, D. P., 2000. Competence, relatedness, and autonomy in life stories. *Psychological Inquiry*, 11(4), pp. 276–279.

Björk, S. & Holopainen, J., 2005. *Patterns in Game Design.* Hingham, MA: Charles River Media.

Bloom, B. S., 1977. Affective outcomes of school learning. *The Phi Delta Kappan,* 59(3), pp. 193–198.

Bødker, S., 1991. *Through the Interface: A Human Activity Approach To User Interface Design.* Mahwah, NJ: Lawrence Erlbaum.

Brito, P. F. d., Moreno, A. D., de Souza, G. B. & Brandão, J. H., 2023. *Learning Object as a Mediator in the User/Learner's Zone of Proximal Development.* Cham: Springer, pp. 285–293.

Brown, L. V., 2007. *Psychology of Motivation.* New York, NY: Nova Publishers.

Burke, B., 2014. *Gamify: How Gamification Motivates People to Do Extraordinary Things.* New York, NY: Routledge.

Cerasoli, C. P., Nickling, J. M. & Nassrelgrgawi, A. S., 2016. Performance, incentives, and needs for autonomy, competence, and relatedness: A meta-analysis. *Motivation and Emotion,* 40, pp. 781–813.

Chishti, Z., 2020. *Gamification Marketing For Dummies.* Hoboken, NJ: Wiley.

Chou, Y.-K., 2016. *Actionable Gamification: Beyond Points, Badges, and Leaderboards.* Fremont, CA: Octalysis Media.

Clark, D., 2021. *Learning Experience Design: How to Create Effective Learning that Works.* London, UK: Kogan Page Publishers.

Conway, S., 2014. Zombification?: Gamification, motivation, and the user. *Journal of Gaming & Virtual Worlds*, 6(2), pp. 129–141.

Coonradt, C. A., 2012. *The Game of Work.* Layton, UT: Gibbs Smith.

Costello, R., 2020. *Gamification Strategies for Retention, Motivation, and Engagement in Higher Education: Emerging Research and*

Opportunities: Emerging Research and Opportunities. Hershey, PA: IGI Global.

Costikyan, G., 2002. *I Have No Words & I Must Design: Toward a Critical Vocabulary of Games.* Tampere, FI: Tampere University Press, pp. 9–33.

Csikszentmihalyi, M., 1991. *Flow: The Psychology of Optimal Experience.* New York, NY: Harper Perennial.

Deterding, S., 2014. Eudaimonic Design, or: Six Invitations to Rethink Gamification. In: *Rethinking Gamification.* Lüneburg, DE: Meson Press, pp. 305–323.

Deterding, S., Dixon, D., Khaled, R. & Nacke, L., 2011. *From Game Design Elements to Gamefulness.* New York, NY: Association for Computing Machinery, pp. 9–15.

Deutscher, I., 1951–1952. Socialist competition. *Foreign Affairs,* 30, pp. 376–390.

Diefenbach, S. & Müssig, A., 2019. Counterproductive effects of gamification: An analysis on the example of the gamified task manager Habitica. *International Journal of Human-Computer Studies,* 127, pp. 190–210.

Djaouti, D., Alvarez, J., Jessel, J.-P. & Rampnoux, O., 2011. Origins of Serious Games. In: *Serious Games and Edutainment Applications.* London, UK: Springer, pp. 24–43.

Donovan, T., 2010. *Replay: The History of Video Games.* East Sussex, UK: Yellow Ant.

Draper, S. W., 1999. Analysing fun as a candidate software requirement. *Personal Technologies,* 3, pp. 117–122.

Egenfeldt-Nielsen, S., Smith, J. H. & Tosca, S. P., 2016. *Understanding Video Games: The Essential Introduction.* Third ed. New York, NY: Routledge.

Ford, M. E. & Smith, P. R., 2020. *Motivating Self and Others: Thriving with Social Purpose, Life Meaning, and the Pursuit of Core Personal Goals.* Cambridge, UK: Cambridge University Press.

Fullerton, T., 2018. *Game Design Workshop: A Playcentric Approach to Creating Innovative Games.* Fourth ed. Boca Raton, FL: CRC Press.

Goethe, O., 2019. *Gamification Mindset.* Cham, Switzerland: Springer.

Grudin, J., 1990. *The Computer Reaches Out: The Historical Continuity of Interface Design.* Seattle, WA: ACM.

Gurjanow, I. et al., 2019. Mathematics trails: Shallow and deep gamification. *International Journal of Serious Games,* 6(3), pp. 65–79motivation.

Hennesey, M. B., Altringer, B. & Amabile, T. M., 2015. Extrinsic and Intrinsic Motivation. In: *Wiley Encyclopedia of Management.* Hoboken, NJ: Wiley, pp. 1–4.

Hilgers, P. V., 2012. *War Games: A History of War on Paper.* Cambridge, MA: The MIT Press.

Hugos, M., 2012. *Enterprise Games: Using Game Mechanics to Build a Better Business.* Sebastopol, CA: O'Reilly Media.

Hulsey, N., 2019. *Games in Everyday Life: For Play.* Bingley, UK: Emerald Publishing.

Hunicke, R., LeBlanc, M. & Zubeck, R., 2004. *MDA: A Formal Approach to Game Design and Game Research.* San Jose, CA: AAAI Press, pp. 2–5.

Jahn, K. et al., 2021. Individualized gamification elements: The impact of avatar and feedback design on reuse intention. *Computers in Human Behavior*, 119(2), pp. 1–13.

Jeal, T., 1989. *Baden-Powell: Founder of the Boy Scouts.* London, UK: Hutchinson.

Jozani, M. M., Maasberg, M. & Ayaburi, E., 2018. *Slayers vs Slackers: An Examination of Users' Competitive Differences in Gamified IT Platforms Based on Hedonic Motivation System Model.* Cham, CH: Springer International Publishing, pp. 164–172.

Kapp, K. M., 2013. *The Gamification of Learning and Instruction Fieldbook: Ideas into Practice.* San Fransisco, CA: Wiley.

Kim, S., Song, K., Lockee, B. & Burton, J., 2017. *Gamification in Learning and Education: Enjoy Learning Like Gaming.* Spring: Cham, CH.

Kocadere, S. A. & Çağlar, Ş. Ç., 2018. Gamification from player type perspective: A case study. *Educational Technology & Society*, 21(3), pp. 12–22.

Korine, H., 2022. *Preventing Corporate Governance Failure.* Bern, CH: Haupt Verlag.

Laurel, B., 2013. *Computers as Theatre.* Second ed. Upper Saddle River, NJ: Addison-Wesley.

Lecture, J. S., 2002. Psychological foundations of incentives. *European Economic Review,* 46(4–5), pp. 687–724.

Lieberoth, A., Møller, M. & Marin, A. C., 2015. Deep and Shallow Gamification in Marketing: Thin Evidence and the Forgotten Powers of Really Good Games. In: *Engaging Consumers through Branded Entertainment and Convergent Media.* Hershey, PA: Business Science Reference, pp. 110–126.

Locke, E. A., 1976. Motivation through conscious goal setting. *Applied and Preventive Psychology,* 5(2), pp. 117–124.

Lunenburg, F. C., 2011. Goal-setting theory of motivation. *International Journal of management, Business, and Administration,* 15(1), pp. 1–6.

Mak, J., 2003. *Tourism and the Economy.* Honolulu, HI: University of Hawaii Press.

Malone, T. W., 1980. *What Makes Things Fun to Learn? Heuristics for Designing Instructional Computer Games.* New York, NY, Association for Computing Machinery, pp. 162–169.

Malone, T. W., 1982. *Heuristics for Designing Enjoyable User Interfaces: Lessons from Computer Games.* New York, NY, Association for Computing Machinery, pp. 63–68.

Manheim, D., 2023. Building less-flawed metrics: Understanding and creating better measurement and incentive systems. *Patterns,* 4(10), pp. 1–12.

Marcusson, L., 2020. Gamification Is. In: *Utilizing Gamification in Servicescapes for Improved Consumer Engagement.* Hershey, PA: IGI Global, pp. 24–51.

Martínez-López, F. J., Li, Y. & Young, S. M., 2022. *Social Media Monetization: Platforms, Strategic Models and Critical Success Factors.* Cham, CH: Springer.

Mavroeidi, A.-G., Kitsiou, A., Kalloniatis, C. & Gritzalis, S., 2019. Gamification vs. privacy: Identifying and analysing the major concerns. *Future Internet,* 11(3), pp. 67–83.

Mazur-Stommen, S. & Farley, K., 2016. *Games for Grownups: The Role of Gamification in Climate Change and Sustainability*. Washington, DC: Indicia Conulting.

Moore, M. E., 2011. *Basics of Game Design.* Boca Raton, FL: CRC Press.

Mozelius, P., 2021. *Deep and Shallow Gamification in Higher Education, What is the Difference?.* Valencia, ES: IATED, pp. 3150–3156.

Nelson, M. J., 2012. *Soviet and American Precursors to the Gamification of Work.* New York, NY: Association for Computing Machinery, pp. 23–26.

Nicholson, S., 2012. *Strategies for Meaningful Gamification: Concepts Behind Transformative Play and Participatory Museums.* Lansing, MI: Michigan State University, pp. 1–16.

Nicholson, S., 2014. Exploring the Endgame of Gamification. In: *Rethinking Gamification.* Lüneburg, DE: Meson Press, pp. 289–304.

Nicholson, S., 2015. A RECIPE for Meaningful Gamification . In: *Gamification in Education and Business.* New York, NY: Springer, pp. 1–20.

Parker, G. T., 1977. How to win friends and influence people: Dale Carnegie and the problem of sincerity. *American Quarterly,* 29(3), pp. 506–518.

Pollack, M., 1988. A company study green stamps: A case study. *Journal of Services Marketing,* 2(4), pp. 37–40.

Prestopnik, N., Crowston, K. & Wang, J., 2017. Gamers, citizen scientists, and data: Exploring participant contributions in two games with a purpose. *Computers in Human Behavior*, 68, pp. 254–268.

Pritchard, A. & Woollard, J., 2013. *Psychology for the Classroom: Constructivism and Social Learning.* New York, NY: Routledge.

Qayum, M., Sawal, S. H. & Khan, H. M., 2014. Motivating employees through incentives: productive or a counterproductive strategy. *Journal of Pakistan Medical Association,* 64(5), pp. 567–570.

Reiss, S., 2012. Intrinsic and extrinsic motivation. *Teaching Psychology,* 39(2), pp. 152–156.

Rieber, R., 1983. *Dialogues on the Psychology of Language and Thought.* New York, NY: Plenum Press .

Risso, M. & Paesano, A., 2021. Retail and gamification for a new customer experience in omnichannel environment. *Journal of Economic Behavior,* 11(1), pp. 109–128.

Romero, M., 2020. *Digital Games and Learning.* Montréal, QC: Editions JFD.

Rudmann, D., 2017. *Learning and Memory.* Thousand Oaks, CA: SAGE Publications.

Sailer, M., Hense, J., Mandl, H. & Klevers, M., 2013. Psychological perspectives on motivation through gamification. *Interaction Design and Architecture(s) Journal,* 19, pp. 28–37.

Sánchez, G. P., Cózar-Gutiérrez, R., Olmo-Muñoz, J. d. & González-Calero, J. A., 2021. Impact of a gamified platform in the promotion of reading comprehension and attitudes towards reading in primary education. Computer Assisted Language Learning, pp. 1–25.

Santos, P. A., 2015. *Deep Gamification of a University Course.* Coimbra, PT: Departamento de Engenharia Electrotécnica Polo II da Universidade de Coimbra, pp. 1–5.

Siebert, H., 2001. *Der Kobra-Effekt: Wie man Irrwege der Wirtschaftspolitik vermeidet.* Munich, DE: Deutsche Verlags-Anstalt.

Skinner, B. F., 1953. *Science and Human Behavior.* New York, NY: Macmillan.

Smith, D., 1998. *Disney A to Z: The Updated Official Encyclopedia.* New York, NY: Hyperion.

Strodder, C., 2012 . *The Disneyland Encyclopedia: The Unofficial, Unauthorized, and Unprecedented History of Every Land, Attraction, Restaurant, Shop, and Major Event in the Original Magic Kingdom.* Solana Beach, CA: Santa Monica Press.

Söbke, H., 2019. *A Case Study of Deep Gamification in Higher Engineering Education.* Cham, CH: Springer International Publishing, pp. 375–386.

Söbke, H. & Londong, J., 2019. *Towards Integration of Deep Gamification Into Formal Educational Settings.* Kidmore End, UK: Academic Conferences International Limited, pp. 519–525.

Toda, A. M. et al., 2019. *A Taxonomy of Game Elements for Gamification in Educational Contexts: Proposal and Evaluation.* Maceio, Brazil: IEEE, pp. 84–88.

Ulrich, D., 1989. Tie the corporate knot: Gaining complete customer commitment. *MIT Sloan Management Review*, 30(4), pp. 19–27.

Vesa, M., 2021. *Organizational Gamification: Theories and Practices of Ludified Work in Late Modernity.* New York, NY: Routledge.

Visser, W., 2006. *The Cognitive Artifacts of Designing.* Mahwah, NJ: Lawrence Erlbaum Associates.

Werbach, K. & Hunter, D., 2012. *For the Win: How Game Thinking Can Revolutionize Your Business.* Philadelphia, PA: Wharton Digital Press.

Wilkinson, P., 1969. English youth movements, 1908–30. *Contemporary History,* 4(2), pp. 3–23.

Wintjes, J., 2016. Not an ordinary game, but a school of war: Notes on the early history of the Prusso-German Kriegsspiel. *Vulcan,* 4, pp. 52–75.

Zicherman, G. & Cunningham, C., 2011. *Gamification by Design: Implementing Game Mechanics in Web and Mobile Apps.* Sebastopol, CA: O'Reilly Media.

2 Industry 4.0 and smart manufacturing

Industrial revolutions

To understand Industry 4.0, it is necessary to recognize the three industrial revolutions that came before. Industry 4.0 is the latest in the succession of industrial revolutions and could not have worked if these revolutions had not paved the way for advanced information and communication technology.

The Industrial Revolution stands as a critical epoch in human history, outlined by a sequence of characteristic stages, each characterized by its own technological breakthroughs and momentous societal transformations. The advent of the Industrial Revolution fundamentally altered the trajectory of civilizations, leaving a permanent mark on the course of our collective future.

Around 1750–1760, the First Industrial Revolution began in England and lasted until 1820–1840 (Mohajan, 2019). During this time, both human and animal labor was used to create technology. Production techniques based on machines were created at this time, doing away with the necessity for labor-intensive human work. Steam engines and textile mills, two revolutionary technological innovations, changed the way goods were produced and transported. The steam engine, the spinning Jenny, a multi-spindle spinning frame, and melting iron ore to get pure iron (smelting), stirring the melted iron to remove impurities (puddling), and shaping the iron by passing it through rollers (rolling coke) are a just a few examples of the ingenuity. The world was changing with things like trains, canned food, and mass media in the form of newspapers. All these marvelous inventions affected society's very foundation. The adoption of these advances increased output and productivity, laying the groundwork for an extended period

DOI: 10.4324/9781003406822-2

of economic expansion and urbanization. The factory system, as well as the emergence of a middle class, were among the epoch-defining outcomes of this era of progress (Deane, 1979; Kovarik, 2011; Loy, et al., 2022).

The technological advances of the Second Industrial Revolution (1867–1914) greatly impacted manufacturing and production, leading to a significant industry-wide transition (Smil, 2005). Decisive advancements such as the assembly line, electricity, and the internal combustion engine wrought profound changes in the ways that production, transportation, and communication were conducted with cars and airplanes, as well as telephone contact and radio transmission. One important innovation was Taylorism, also known as scientific management, which is a theory of management that analyzes and optimizes workflows to improve labor productivity, developed by Frederick Winslow Taylor in the early 20th century (Littler, 1978). The far-reaching effects of the Second Industrial Revolution extended beyond industry, with significant societal impacts that included the creation of fresh employment prospects, the promotion of urban expansion, and the ongoing evolution of social and economic frameworks, such as the ascent of consumer culture and the emergence of novel modes of mass communication and entertainment (Kovarik, 2011; Coluccia, 2012; Meller, 2013; Loy, et al., 2022).

Yet another technological shift took place in the latter half of the twentieth century. This transformation is widely recognized as the Third Industrial Revolution or Digital Revolution, a period of widespread digitization that completely overhauled industrial production and the fabric of society itself (Poster, 1990; Hassan, 2008). Work organizations changed, shifting from physical labor to knowledge-based employment. Innovative automation technologies emerged, made possible by computer usage, such as robots, computer networking, and software development. These advancements significantly impacted society, giving rise to new industries such as telecommunications, e-commerce, and software development. The extensive use of computers and the automation of numerous tasks built the groundwork for a global economy that transcended geographical boundaries (Greenwood, 1997; Fuchs, 2008; Brynjolfsson & McAfee, 2014).

Industry 4.0 and smart manufacturing are the names of yet another technological revolution that has been occurring in the industrial and manufacturing sector since 2011 (Nayya & Kumar, 2019; Tarantino, 2022). By fusing cyber-physical systems, the Internet of Things, augmented reality, and artificial intelligence, this era ushers in a new dawn by building on the digital technology and automation of the Third Industrial Revolution. The result is a network of connected, flexible factories that are very productive, inventive, and efficient (Schwab, 2017). Industry 4.0, the current epoch of the Industrial Revolution, has not only caused significant disruptions to traditional business models but has also presented unique opportunities for sustainability and expansion (Parker, et al., 2016). However, it has brought along several challenges, including retraining of the workforce, safeguarding cybersecurity, and addressing ethical issues concerning privacy. The anticipated impact of this revolution is expected to have effects beyond commerce, with implications for various aspects of human life, work, and social interactions, ultimately resulting in a transformative impact on the global society and economy (Schwab, 2017; Plonka, et al., 2022).

One could count the Industrial Revolutions as either one industrial revolution with four different phases or four distinct industrial revolutions taking place one after another. The initial two industrial revolutions brought about a significant change, transitioning from an agricultural-based society that had existed for a span of ten millennia to an industrial society wherein the primary workforce was engaged in manual labor within the realm of industrial production. The latter two industrial revolutions witnessed the change from an industrial society to an information and service society in which the workforce shifted from industrial labor towards knowledge-based employment but also many service jobs arose. It is important to acknowledge that what is commonly known as information society is about not only the expansion of knowledge work but also many recently emerged service professions (Webster, 2014). Furthermore, it must be emphasized that the world production of agricultural and industrial goods is massive, but in the Western world, at least, it is just a minority of the population working within these fields due to outsourcing, machines, and automated mass production. This means that there is a surplus for society as a whole to be working with other activities than agriculture and industrial production (Stearns, 2020).

Smart manufacturing and Industry 4.0

Industry 4.0, the fourth industrial revolution, merges advanced technologies such as AI, big data, and the Internet of Things to streamline and automate production processes, offering businesses novel opportunities to enhance operational efficiency (Schwab, 2017). The smart manufacturing of Industry 4.0 initiatives uses these technologies to boost output, reduce waste, and improve productivity. Using sensors, data analytics, and automation, smart manufacturing enables real-time supervision and administration of the production process, ultimately enhancing quality and elevating productivity. Industries across the board are embracing smart manufacturing and acknowledging its role as a pivotal driver of competitiveness and economic progress (Pascual, et al., 2019).

In this way, Industry 4.0 is developing the manufacturing sector. It symbolizes a significant transition in the way things are made by fusing cutting-edge computerization, automation, and data exchange with traditional manufacturing processes. Industry 4.0 creates a highly automated and networked production system that can manufacture items with unmatched efficiency, precision, and perfection by fusing smart technologies, advanced data analytics, and connectivity. This revolutionary change in manufacturing processes heralds a new era of production that combines technology and inventiveness to produce a creative and dynamic production ecosystem (Schwab, 2017; Pascual, et al., 2019; Plonka, et al., 2022).

The German government launched the Industry 4.0 program in 2011, which aims to combine new digital technologies with established industrial practices (Chiarello, et al., 2020). The resulting smart factories are able to communicate, exchange data, review it in real-time, and make intelligent judgments, which leads to a beneficial process that is also good for the environment. Because of this paradigm shift, businesses now have a unique chance to sharpen their competitive edge and preserve their dominance in the fast-developing digital economy (Pascual, et al., 2019). The impact of these technical advancements is altering the manufacturing sector (Sadiku, et al., 2023). The implementation of Industry 4.0 marks the commencement of a novel period of innovation and progress that has the potential to significantly alter our professional and personal lifestyles.

The fourth industrial revolution, or Industry 4.0, heralds an outstanding approach to manufacturing goods by utilizing state-of-the-art technology to establish a flexible, efficient, and productive production landscape. The keystone of this methodology is smart manufacturing, which facilitates prompt data collection, analysis, and decision-making by interlinking equipment, devices, and systems. Responses to manufacturing challenges become quicker and more precise as a result. The advantages are numerous, including enhanced quality control, decreased waste, and more chances for customization, allowing companies to create products that are specially tailored for each client. Smart manufacturing represents a groundbreaking transformation in the manufacturing domain, empowering enterprises to uphold competitiveness in the face of ever-evolving market dynamics by optimizing operations and enhancing efficacy (Garbie & Parsaei, 2022).

The growth of the industrial sector right now depends on Industry 4.0 and smart manufacturing. Industry 4.0 paves the way for intelligent manufacturing by making it possible to create a networked, automated, and intelligent production system through the use of pioneering tools and extensive data analysis. Overall productivity, efficiency, and quality need to be increased as a result of this thorough revamp of the production process (Wolf & Lepratti, 2020; Garbie & Parsaei, 2022).

Because it can collect and analyze large amounts of data throughout the manufacturing process, smart manufacturing is very valuable. Manufacturers may improve decision-making, operational efficiency, and product quality by using this data-centric strategy (Cheng, et al., 2017). Smart manufacturing also makes use of automation and robots to offer a flexible and dynamic production process that can react quickly to changes in demand or product design (Evjemo, et al., 2020; Gilberti, et al., 2022; Namjoshi & Manish, 2022). Human workers are freed to concentrate on more complex jobs thanks to this flexibility, which again boosts output and efficiency (Longo, et al., 2017).

Some of the key technologies used in Industry 4.0 and smart manufacturing are as follows:

- Industrial Control Systems (ICS)
- Cyber-Physical Systems (CPS)
- Internet of Things (IoT) – especially Industrial Internet of Things (IIoT)
- Artificial Intelligence (AI)
- Big Data
- Augmented Reality (AR)
- Virtual Reality (VR)

- Cloud Computing
- Digital Twin

It is critical to understand that Industry 4.0 builds on past technological revolutions and significantly improves on Third-Industrial Revolution-era technologies in terms of computerization and automation. This indicates that Industry 4.0 still relies heavily on computers, robots, and control systems like SCADA (Misra, et al., 2021).

Supervisory control and data acquisition (SCADA)

Industrial control systems (ICS) are essential elements of contemporary infrastructure that allow a variety of industrial processes to operate effectively and dependably. Key processes in industries like energy, manufacturing, transportation, and water treatment are controlled and monitored by these systems. Supervisory Control and Data Acquisition (SCADA) is one of the most common ICS, forming the core of today's industrial infrastructure (Daneels & Salter, 1999; Misra, et al., 2021). This essential technology keeps various industries running smoothly by monitoring and controlling their operations. In order to gather real-time data from sensors and devices, SCADA systems combine hardware and software. This enables operators to remotely monitor and manage industrial processes. SCADA is essential for maximizing productivity, guaranteeing safety, and improving operational efficiency across a variety of sectors due to its versatility in subsystem integration. SCADA is a crucial component that provides unrivaled control and optimization capabilities by leveraging the power of real-time data collecting and processing. This groundbreaking technology has signaled a shift away from traditional modalities of control by empowering decision-making with the aid of data-driven insights. Due to SCADA, which enables technology to function at its peak, data has come to dominate the modern industry (Dey & Sen, 2020; Misra, et al., 2021).

The first generation of SCADA was monolithic and controlled by a single mainframe, the second generation had distributed LAN technology that was able to control multiple systems, and the third generation was networked SCADA, creating an open system architecture. The fourth generation of SCADA systems is building upon the earlier generation, adding the Internet of Things with WAN technology (Dey & Sen, 2020). The newest developments of SCADA put a lot of pressure on network security so the systems are not hacked by outside perpetrators (Shaw, 2021).

A SCADA system usually comprises the following elements in its architecture:

- SCADA Master Station
- Actuators
- Sensors
- Remote Terminal Unit(s) (RTU)
- Programmable Logic Controller(s) (PLC)
- Human-Machine Interface (HMI)
- Historian
- Communication Network

A small system's SCADA Master Station is a single computer in charge of coordinating communications with field devices (Bailey & Wright, 2003). Many servers, remote software applications, and disaster recovery sites make up the master station of a sizable SCADA system. By viewing the acquired data on the master station, the operator can perform remote control actions. Thanks to the precise and timely data, which is often accessible in real-time, plant and process operations can be optimized (see Figure 2.1).

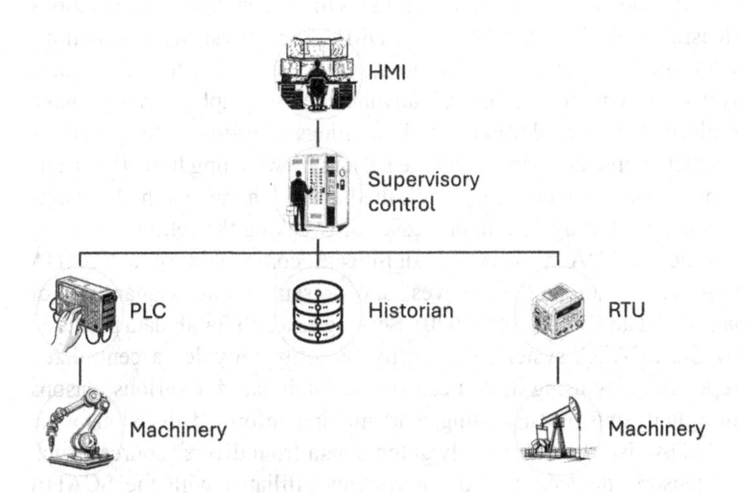

Figure 2.1 SCADA overview.

Sensors are apparatuses that apprehend alterations in the physical or chemical milieu surrounding them and react to these stimuli by transforming them into electric or mechanical impulses. Actuators, on the other hand, are contraptions that regulate or propel a mechanism or system by transmuting the energy into physical motion or force (Eren, 2018; Bolton, 2021).

The fundamental SCADA architecture commences with the utilization of programmable logic controllers (PLCs) or remote terminal units (RTUs) which establish communication channels with the various factory machines, sensors, and actuators. RTUs are in charge of sensing and controlling physical processes at the field level by gathering data from sensors and sending instructions to actuators. Similar to this, PLCs, which are specialized computers, carry out certain logic by directives from the master station and inputs from sensors. The main difference between an RTU (Remote Terminal Unit) and a PLC (Programmable Logic Controller) is that an RTU is designed for remote monitoring and control of field devices in harsh environments, while a PLC is used for automation of industrial processes and is typically installed in a controlled environment. The input-output modules establish connections with both PLCs and RTUs, operating in coordination with the SCADA software system (Strauss, 2003).

The HMI (Human-Machine Interface) holds significant importance in industrial control systems (ICS) as it visually represents the system's operations, empowering users to execute appropriate actions (Misra, et al., 2021). In the past, HMIs were physical constructions with walls covered in indicators, dials, and adjustment buttons. However, with technological advancements, graphic screens have replaced them. In addition to making things comfortable to use, HMIs also let people keep an eye on the system by watching how its various aspects change over time. They can also take charge and make things happen by starting certain processes or tweaking the settings.

The SCADA historian, a significant component of a SCADA system, accumulates, preserves, and presents copious quantities of archival data (Dey & Sen, 2020). Serving as a hub for all data procured by the SCADA system, this software entity provides a centralized repository. By using it, you can sort through data for various reasons like analyzing, documenting, and making informed decisions. The SCADA historian incessantly gathers data from diverse sources, such as sensors, devices, and other systems affiliated with the SCADA network. The data is then stowed away in a database, customarily in a time-series format, that permits facile retrieval and analysis. The historian database can stockpile data for lengthy durations. It is not recommended to simply input unstructured data into the historian system without proper organization. This data can be challenging to interpret and understand. Therefore, it is essential to implement a complex tag naming system to facilitate efficient data storage and retrieval (Greeff & Ghoshal, 2004).

The data amassed in the historian can be employed for diverse objectives such as performance optimization, trend analysis, predictive maintenance, and energy management. It can aid operators and engineers in identifying patterns and trends in the data, spotting irregularities, and making informed determinations based on historical data. The historian database can also be merged with other software applications, such as data analytics and business intelligence tools, to provide more intricate analysis and reporting capabilities. This assimilation enables users to execute complex analyses on vast amounts of data such as predictive modeling, optimization, and machine learning (Borlase, 2018).

SCADA relies on communication networks to establish a vital connection between a central control center and field devices scattered in different locations. These networks serve as the basic framework for exchanging real-time information, enabling smooth monitoring, control, and data collection from sensors, actuators, and controllers. Wired protocols like Ethernet and wireless options such as radio frequency or cellular are used in SCADA systems to ensure secure and reliable data transmission. By utilizing these networks, operators can remotely oversee and manage industrial processes, resulting in enhanced efficiency, responsiveness, and the ability to promptly address abnormalities or emergencies.

Purdue Enterprise Reference Architecture (PERA)

The Purdue Enterprise Reference Architecture was developed in the 1990s to secure cybersecurity for manufacturing companies. The model is also sometimes called the Purdue model, and it was developed by Theodore J. Williams and members of the Industry-Purdue University Consortium for Computer Integrated Manufacturing (Williams, 1994; Li & Williams, 2000; Flaus, 2019). The PERA model operates with several layers to handle cybersecurity information management (CIM).

At the lowest level, Level 0, we have the physical processes, which is where sensors and actuators are located. This is where the operational sequences of the machines take place. So, Level 0 includes the physical processes and the machines and equipment that perform the production tasks, and here we are talking about, for example, motors and pumps. This level varies greatly between different industries, and therefore PERA does not apply at this level.

Level 1: The station level or basic control level has control over production, which involves detection, monitoring, and control of the physical operations with workshop equipment, performed by information processing systems such as programmable logic controllers (PLCs) and Remote Terminal Units (RTUs). It is data collected from sensors and continuous control algorithms that operate at this level. This is about the very basic control functions that directly manage the physical processes at Level 0.

Level 2: Control Systems. This involves supervising, monitoring, and controlling physical processes, which includes the Human-Machine Interface from SCADA or similar systems. At this level, operators interact with the systems to monitor and control the processes.

Level 3: Operations Management. At this level, production is coordinated by managing the operational functions at the factory or plant. Here, one will find manufacturing execution systems (MES) and historical databases. This is where inventory management and quality control are handled.

Level 4: Enterprise Business Systems. This level involves planning tasks, scheduling, and production planning, using IT systems like Enterprise Resource Planning (ERP) that manage and administer business processes such as finance, HR, supply chains, and other business logistics. It is at this level that management relates to production.

With PERA, engineers and programmers gain a framework to structure security in industrial environments with information systems. It also helps in understanding the sensitivity of data a given gamification solution has, depending on how deep into the security levels it is necessary to go to obtain the required information for gamification. This framework ensures that security protocols are appropriately applied and managed, safeguarding sensitive data while enabling the effective implementation of gamification strategies in industrial settings.

Cyber-physical systems and the Industrial Internet of Things

The fusion of the Industrial Internet of Things (IIoT) and cyber-physical systems (CPS) has resulted in the ability to exercise detailed control over industrial machinery, surpassing previous levels observed within the industrial landscape. In their own ways, CPS and IIoT represent the integration of digital technologies with physical processes, enabling improved automation, monitoring, and control (Pivoto, et al., 2021).

CPS refers to the integration of computation, communication, and control into physical systems, creating an interdependent relationship between the cyber and physical components (Putnik, et al., 2019). These systems use embedded sensors, actuators, and processors to collect and communicate physical world data, analyze it in real-time, and produce the necessary reactions as needed. The goal of CPS is to improve the efficiency, reliability, safety, and functionality of physical systems by leveraging advanced computing capabilities (Taha, et al., 2021).

On the other hand, IIoT specifically focuses on the application of Internet of Things (IoT) technologies within industrial settings (Veneri & Capasso, 2018). It encompasses the interlinking of diverse devices, equipment, and systems employed in manufacturing, transportation, energy, and comparable industrial domains. Through the establishment of internet connectivity and data interchange, IIoT facilitates constant surveillance, data scrutiny, and astute decision-making, culminating in enhanced operational efficiency, expenditure reduction, and heightened productivity.

The integration of CPS and IIoT has opened up new possibilities for industrial processes and systems. By deploying sensors and actuators within physical systems, CPS enables the collection of vast amounts of data, ranging from environmental conditions to equipment performance. Data is sent across networks and examined using sophisticated algorithms like machine learning and artificial intelligence (Vermesan, et al., 2020). This process aims to extract meaningful information and facilitate intelligent decision-making (Tahir, et al., 2021).

The benefits of CPS and IIoT are numerous. With real-time monitoring and predictive analytics, potential issues and faults can be identified before they cause significant disruptions or failures. This proactive approach to maintenance, known as predictive maintenance, reduces downtime, extends equipment lifespan, and optimizes maintenance schedules. Additionally, CPS and IIoT enable remote monitoring and control, allowing operators to access and manage systems from anywhere, enhancing operational flexibility and reducing the need for physical presence.

Big data and artificial intelligence

Big data refers to the large volume of complex and diverse data that is generated at a high velocity from various sources. These data include structured and unstructured information, such as sensor readings and transaction records. Big data is characterized by its three key attributes: volume, velocity, and variety (Ghavami, 2020; Tahir, et al., 2021).

Volume: Big data involves vast amounts of data that exceed the processing capabilities of traditional data management systems. It includes data from multiple sources and accumulates at an unprecedented rate.

Velocity: Big data is generated and collected at a high speed. This real-time or near-real-time data flow requires efficient processing and analysis to derive meaningful insights and make timely decisions.

Variety: Big data includes a diverse multitude of data types and configurations, unveiling the presence of structured, semi-structured, and unstructured data. When discussing Industry 4.0, the primary focus is on reading sensors, although it may likewise involve text, images, and other sources. Managing and integrating these diverse data sources can be challenging.

Artificial intelligence is machines, particularly computer systems that simulate human intelligence processes. AI algorithms and techniques facilitate machines in processing and examining substantial amounts of data, including big data, in order to extract patterns, make predictions, and automate tasks. AI techniques commonly applied to big data include (Ghavami, 2020):

1. Machine Learning (ML): ML algorithms allow systems to automatically learn and improve from experience without being explicitly

programmed. They can analyze large datasets to discover patterns, relationships, and anomalies. ML models are trained on historical data and then used to make predictions or classify new data.

2. Deep Learning: Deep Learning is a specialized branch of machine learning that places emphasis on the utilization of neural networks consisting of numerous layers. Its capacity to handle many unstructured data formats is its key strength. Deep Learning algorithms can automatically learn hierarchical representations of data, enabling more sophisticated pattern recognition and analysis. This is used in generative AI such as ChatGPT.
3. Natural Language Processing (NLP): NLP techniques facilitate the comprehension and analysis of human language by computers. They are used to process and extract insights from unstructured textual data, such as social media posts, customer reviews, and documents.
4. Data Mining: Data mining techniques involve discovering patterns and extracting knowledge from large datasets. These techniques help uncover hidden relationships, identify trends, and make predictions.

Organizations can gain useful insights, improve their decision-making procedures, and promote innovation by combining big data and AI. Large amounts of data can be processed and analyzed using AI algorithms, revealing patterns, correlations, and trends that may escape human observation.

Augmented reality (AR) and virtual reality (VR)

Augmented reality (AR), an interactive experience, combines the real environment and computer-generated material. It helps increase productivity and efficiency and has many uses in manufacturing and production. AR is a powerful tool for training and education across various industries. By blending virtual elements with the real world, AR provides a unique and immersive learning experience (Schmalstieg & Höllerer, 2016).

AR revolutionizes remote assistance and collaboration by allowing experts to provide real-time guidance to on-site workers through smart glasses or mobile applications (Anderson, 2019). It improves workplace safety by offering real-time information about hazards and simulating risky scenarios for training purposes. AR enhances the product design and prototyping stages by enabling engineers to visualize and manipulate virtual 3D models in a real-world context, saving time and costs. It also improves maintenance and repair processes by overlaying digital instructions and sensor data onto physical equipment, reducing downtime. In logistics and warehousing, AR provides real-time information on inventory and optimal routes, streamlining operations. AR is also valuable in quality control and inspection, allowing inspectors to identify defects and deviations more efficiently. Additionally, AR facilitates data visualization and analytics by overlaying intuitive visualizations onto physical environments, aiding decision-making based on real-time data (Aukstakalnis, 2016).

Virtual reality (VR), like augmented reality (AR), provides a technology-supported experience. However, while augmented reality offers it as an extra layer to the surrounding reality, virtual reality becomes an all-encompassing simulated reality that the user can step into and immerse themselves in (Wohlgenannt, et al., 2020). It gives the experience of presence in another reality than the one the user otherwise finds themselves in. However, not all senses are necessarily affected. In many cases, VR primarily affects the visual and auditory senses, but it can also somewhat affect the sense of touch and possibly even gravity. In principle, all senses can be affected, thereby creating a realistic simulation. It should be noted that the more senses that need to be affected, the more expensive the solution becomes.

Virtual reality has gained momentum with great success in the training of pilots and other large machinery (Grabowski, 2020). Here, the advantage is that it is much cheaper to crash a simulation of a plane than to crash a plane with passengers. For the same reason, there have been great expectations for the use of virtual reality in Industry 4.0. The reason it has not yet fully taken off is that the equipment is still relatively expensive, the experience can cause dizziness and motion sickness, the software and machinery may be expensive to update if routines are changed, and it needs to have a pedagogical purpose rather than a distraction (Yildirim, 2019; Lege & Bonner, 2020). Nevertheless, research in VR continues, and solutions have also become cheaper over time. However, one should be careful not to be carried away by the hype but, instead, focus sharply on how AR and VR can make a positive difference in the industry (Mealy, 2018).

Cloud computing

The central position occupied by cloud computing within the Industry 4.0 paradigm emerges from its provision of indispensable infrastructure for manufacturers, facilitating the storage, management, and processing of extensive volumes of data. By capitalizing on remote servers hosted on the internet, cloud computing offers manufacturers and entrepreneurs a vast selection of software programs, endowed with unprecedented operational efficiency, scalability, and flexibility (Dubey, 2018).

The magnitude of cloud computing's importance within the framework of Industry 4.0 should not be underestimated, as its impact is profound (Majstorovic & Stojadinovic, 2020). Manufacturers, as prolific generators of data stemming from miscellaneous sources including sensors, machines, IoT devices, cyber-physical systems, and production lines, benefit from the secure storage and ubiquitous accessibility afforded by cloud computing. Real-time decision-making and analysis are facilitated by the availability and accessibility of data, leading to improved operational efficiency. Additionally, cloud computing facilitates the remote management of software programs by manufacturers, thus eliminating the requirement for locally installed applications. This transformative shift renders elaborate IT infrastructure unnecessary and leads to diminished maintenance expenses.

Manufacturers possess the ability to conveniently modify their software utilization to align with their specific requirements, thereby augmenting their operational adaptability and cost-efficiency. What is more, the integration of cloud-based platforms and tools facilitates seamless collaboration and communication amongst manufacturers, suppliers, partners, and customers. This synergetic environment nurtures streamlined supply chain management, accelerates innovation, and facilitates prompt responsiveness to market demands. Actionable insights may be extracted from large datasets, processes can be automated, and operations can be optimized by combining these technologies with cloud platforms. Predictive analytics enabled by cloud-based AI and ML models efficiently analyze large amounts of data, enabling proactive maintenance, quality control, and resource management (Nath, et al., 2020; Mourtzis, 2021).

Digital twin

Digital twin technology combines advanced sensor networks, machine learning algorithms, and simulation software to produce a virtual replica that accurately represents a physical system (VanDerHorn & Mahadevan, 2021; Vohra, 2023; Ganguli, 2023). This integration allows for real-time monitoring and thorough analysis of the physical system, providing us with predictive insights and immersive simulations of its future behaviors. This technological advancement has considerably impacted the fields of physical asset design, operation, and maintenance, opening up new possibilities. Engineers and operators now have the opportunity to explore asset performance, uncover patterns, and optimize relevant areas. With digital twin technology, efficiency, and productivity improve as understanding of complex systems reaches unprecedented levels.

Additionally, digital twin technology has the capability to replicate and predict future situations, empowering operators to make well-informed choices regarding their response to dynamic circumstances and imaginable disturbances. By modeling various scenarios and analyzing outcomes, downtime may be minimized, and the lifespan of assets can be extended. The digital twin's forecasting capabilities have significant implications across industries, and its potential impact is only beginning to be realized (Ganguli, 2023).

Digital twin technology is an integral part of Industry 4.0 and smart manufacturing, allowing machines and equipment to communicate and exchange data seamlessly (Zidek, et al., 2020; Moiceanu & Paraschiv, 2022). By creating virtual replicas of these machines, manufacturers can simulate and predict outcomes, which supposedly improves efficiency and decision-making. The continuous monitoring and analysis of sensor data is said to help detect potential issues early, leading to less downtime and reduced maintenance expenses. Digital twins also aid in simulating production processes for analysis, identifying bottlenecks, and optimizing performance. Thereby supposedly enhancing operational efficiency and productivity while reducing waste.

Gamification of Industry 4.0 and smart manufacturing

It is crucial for a project leader and a game designer tasked with creating gamification for the industry to understand how smart factories and Industry 4.0 operate in practice. Having a profound understanding of game mechanics alone is insufficient without a fundamental grasp of how a factory functions. Therefore, the project leader and the game designer must comprehend an Industrial Control System (ICS), such as SCADA, and relate it to the Industrial Internet of Things (IIoT) and cyber-physical systems (CPS). Additionally, utilizing cloud computing, augmented reality (AR), virtual reality (VR), and digital twin in the context of gamifying industrial processes is possible.

The project leader and the game must obtain the competence to grasp industrial control systems (ICS). It becomes paramount to comprehend the correlations and interconnectedness of the many elements of the industrial manufacturing processes, enabling the designer to develop gamification systems that promote and encourage efficiency combined with the emotional satisfaction of doing the tasks. When a game designer practices his game mechanics skills and links this to knowledge about the functionality of Industry 4.0 and smart manufacturing, the designer gains the capacity to create gamification experiences with meaningful impact. Thereby not only increasing productivity but also improving quality of work-life.

Now, before we move on to figuring out how a project leader and a game designer develop functional gamification solutions that promote productivity, motivation, and learning, we will look at how the industry can use gamification. For a game designer, it is less important whether it is gamification or serious games. There is a difference in how it is developed and implemented, but from a project leader and a game designer's perspective, the most important thing is what works most effectively. In the next chapter, we will look at various opportunities for gamifying smart manufacturing in Industry 4.0.

Summary

The three preceding industrial revolutions have cumulatively led to the advent of the current epoch, termed Industry 4.0. This revolution is characterized by the pervasive utilization of electronic networks, artificial intelligence, and cyber-physical systems. Integral technologies such as smart robotics, drones, virtual reality (VR), and augmented reality (AR) constitute some of the multifarious opportunities available for leveraging the Industrial Internet of Things (IIoT) within smart factories.

In the current industrial paradigm, considerable importance is ascribed to the implementation of digital twins, big data, and cloud computing, all of which facilitate the optimization and automation of production processes, in addition to enhancing data-driven decision-making. A quintessential element of this progression is the SCADA (Supervisory Control and Data Acquisition) systems or analogous industrial control systems. To efficaciously implement and capitalize on SCADA systems, it is imperative to possess a profound comprehension of their functionalities. Concurrently, it is indispensable to understand the various security levels delineated in the PERA (Purdue Enterprise Reference Architecture) model to ensure the effective and secure operation of smart factories.

Bibliography

Anderson, A., 2019. *Virtual Reality, Augmented Reality and Artificial Intelligence in Special Education: A Practical Guide to Supporting Students with Learning Differences.* New York, NY: Routledge.

Aukstakalnis, S., 2016. *Practical Augmented Reality: A Guide to the Technologies, Applications, and Human Factors for AR and VR.* Boston, MA: Addison-Wesley.

Bailey, D. & Wright, E., 2003. *Practical SCADA for Industry.* Burlington, MA: Elsevier.

Bolton, W., 2021. *Instrumentation and Control Systems*. third ed. Cambridge, MA: Newnes.

Borlase, S., 2018. *Smart Grids: Advanced Technologies and Solutions*. Second Ed. Boca Raton, FL: CRC Press.

Brynjolfsson, E. & McAfee, A., 2014. *The Second Machine Age: Work, Progress, and Prosperity in a Time of Brilliant Technologies.* New York, NY: W. W. Norton & Company.

Cheng, B. et al., 2017. Smart factory of industry 4.0: Key technologies, application case, and challenges. *IEEE Explore*, IEEE.

Chiarello, F., Trivelli, L., Bonaccorsi, A. & Fantoni, G., 2020. Behind the definition of industry 4.0: Analysis and open questions. *International Journal of Production Economics*, 226, pp. 244–257.

Coluccia, D., 2012. The Second Industrial Revolution (late 1800s and early 1900s). In: *Corporate Management in a Knowledge-Based Economy.* New York, NY: Palgrave-Macmillan, pp. 52–64.

Daneels, A. & Salter, W., 1999. *What is SCADA?.* Trieste, IT: CERN, pp. 339–343.

Deane, P., 1979. *The First Industrial Revolution.* Second ed. Cambridge, UK: Cambridge University Press.

Dey, C. & Sen, S. K., 2020. *Industrial Automation Technologies.* Boca Raton, FL: CRC Press.

Dubey, S. S., 2018. *Cloud Computing and Beyond: A Managerial Perspective.* Second ed. New Delhi, IN: I.K: International Publishing House.

Eren, H., 2018. *Wireless Sensors and Instruments: Networks, Design, and Applications.* Boca Raton, FL: CRC Press.

Evjemo, L. D., Gjerstad, T., Grøtli, E. I. & Sziebig, G., 2020. Trends in smart manufacturing: Role of humans and industrial robots in smart factories. *Current Robotics Reports,* 1, pp. 35–41.

Flaus, J.-M., 2019. *Cybersecurity of Industrial Systems.* Hoboken, NJ: Wiley.

Fuchs, C., 2008. *Internet and Society: Social Theory in the Information Age.* New York, NY: Routledge.

Ganguli, R., 2023. *Digital Twin: A Dynamic System and Computing Perspective.* Boca Raton, FL: CRC Press.

Garbie, I. H. & Parsaei, H. R., 2022. *Reconfigurable Manufacturing Enterprises for Industry 4.0.* Boca Raton, FL: CRC Press.

Ghavami, P., 2020. *Big Data Analytics Methods: Analytics Techniques in Data Mining, Deep Learning and Natural Language Processing.* second ed. Berlin, D&D: Walter de Gruyter.

Gilberti, H. et al., 2022. A methodology for flexible implementation of collaborative robots in smart manufacturing systems. *Robotics,* 11(1), pp. 1–13.

Grabowski, A., 2020. *Virtual Reality and Virtual Environments: A Tool for Improving Occupational Safety and Health.* Boca Raton, FL: CRC Press.

Greeff, G. & Ghoshal, R., 2004. *Practical E-Manufacturing and Supply Chain Management.* Burlington, MA: Elsevier.

Greenwood, J., 1997. *The Third Industrial Revolution: Technology, Productivity, and Income Inequality.* Washington, DC: The AEI Press.

Hassan, R., 2008. *The Information Society: Cyber Dreams and Digital Nightmares.* Cambridge, UK: Polity Press.

Kovarik, B., 2011. *Revolutions in Communications: Media History from Gutenberg to the Digital Age.* New York, NY: Continuum.

Lege, R. & Bonner, E., 2020. Virtual reality in education: The promise, progress, and challenge. *JALT CALL Journal,* 16(3), pp. 167–180.

Li, H. & Williams, T. J., 2000. The interconnected chain of enterprises as presented by the Purdue Enterprise Reference Architecture. *Computers in Industry,* 42(2–3), pp. 265–274.

Littler, C. R., 1978. Understanding Taylorism. *British Journal of Sociology,* 29(2), pp. 185–202.

Longo, F., Nicoletti, L. & Padovano, A., 2017. Smart operators in industry 4.0: A human-centered approach to enhance operators' capabilities and competencies within the new smart factory context. *Computers & Industrial Engineering,* 113, pp. 144–159.

Loy, A. C. M., Chin, B. L. & Sankaran, R., 2022. Industrial Revolution 1.0 and 2.0. In: *The Prospect of Industry 5.0 in Biomanufacturing.* Boca Raton, FL: CRC Press, pp. 1–14.

Majstorovic, V. D. & Stojadinovic, S. M., 2020. Cloud Computing: Virtualization, Simulation and Cybersecurity – Cloud Manufacturing Issue. In: *Enabling Technologies for the Successful Deployment of Industry 4.0.* Boca Raton, FL: CRC Press, pp. 85–104.

Mealy, P., 2018. *Virtual & Augmented Reality For Dummies.* Hoboken, NJ: Wiley.

Meller, H., 2013. *Leisure and the Changing City 1870–1914.* New York, NY: Routledge.

Misra, S., Roy, C. & Mukherjee, A., 2021. *Introduction to Industrial Internet of Things and Industry 4.0.* Boca Raton, FL: CRC Press.

Mohajan, H., 2019. The first industrial revolution: Creation of a new global human era. *Journal of Social Sciences and Humanities*, 5(4), pp. 377–387.

Moiceanu, G. & Paraschiv, G., 2022. Digital twin and smart manufacturing in industries: A bibliometric analysis with a focus on industry 4.0. *Sensors*, 22(4), pp. 1388–1409.

Mourtzis, D., 2021. *Design and Operation of Production Networks for Mass Personalization in the Era of Cloud Technology.* Cambridge, MA: Elsevier.

Namjoshi, J. & Manish, R., 2022. Role of smart manufacturing in industry 4.0. *Materials Today: Proceedings*, 63, pp. 475–478.

Nath, S. V., Dunkin, A., Chowdhary, M. & Patel, N., 2020. *Industrial Digital Transformation: Accelerate Digital Transformation with Business Optimization, AI, and Industry 4.0.* Birmingham, UK: Packt Publishing.

Nayya, A. & Kumar, A., 2019. *A Roadmap to Industry 4.0: Smart Production, Sharp Business and Sustainable Development.* Cham, CH: Springer.

Parker, G. G., Alstyne, M. W. & Choudary, S. P., 2016. *Platform Revolution: How Networked Markets are Transforming the Economy and How to Make Them Work for You.* New York, NY: W. W. Norton.

Pascual, D. G., Daponte, P. & Kumar, U., 2019. *Handbook of Industry 4.0 and SMART Systems.* Boca Raton, FL: CRC Press.

Pivoto, D. g. S. et al., 2021. Cyber-physical systems architectures for industrial internet of things applications in industry 4.0: A literature review. *Journal of Manufacturing Systems,* 68(A), pp. 176–192.

Plonka, M., Kożuch, M. & Stanienda, J., 2022. The Fourth Industrial Revolution and Contemporary Technological, Economic and Cultural Megatrends. In: *Public Goods and the Fourth Industrial Revolution: Inclusive Models of Finance, Distribution and Production.* Lonodn, UK: Routledge, pp. 7–34.

Poster, M., 1990. *The Mode of Information: Poststructuralism and Social Contexts.* Cambridge, UK: Polity Press.

Putnik, G. D., Ferreira, L., Lopes, N. & Putnik, Z., 2019. What is a Cyber-Physical System: Definitions and models spectrum. *FME Transactions*, 47(4), pp. 663–674.

Sadiku, M. N. O., Ajayi-Majebi, A. J. & Adebo, P. O., 2023. *Emerging Technologies in Manufacturing.* Cham, CH: Springer.

Schmalstieg, D. & Höllerer, T., 2016. *Augmented Reality: Principles and Practice.* Boston, MA: Addison-Wesley.

Schwab, K., 2017. *The Fourth Industrial Revolution.* New York, NY: Crown.

Shaw, W. T., 2021. *Cybersecurity for SCADA Systems.* Tulsa, OK: PennWell Books.

Smil, V., 2005. *Creating the Twentieth Century: Technical Innovations of 1867–1914 and Their Lasting Impact.* New York, NY: Oxford University Press.

Stearns, P. N., 2020. *The Industrial Revolution in World History.* New York, NY: Routledge.

Strauss, C., 2003. *Practical Electrical Network Automation and Communication Systems.* Burlington, MA: Newnes.

Taha, W. M., Taha, A.-E. M. & Thunberg, J., 2021. *Cyber-Physical Systems: A Model-Based Approach.* Cham, CH: Springer.

Tahir, M., Hamadneh, N. N. & Rahmani, M. K., 2021. Machine Learning and Deep Learning Are Crucial to the Existence of IoT and Big Data. In: *A Step Towards Society 5.0: Research, Innovations, and Developments in Cloud-Based Computing Technologies.* Boca Raton, FL: CRC Press, pp. 69–77.

Tarantino, A., 2022. *Smart Manufacturing: The Lean Six Sigma Way.* Newark, NJ: Wiley.

VanDerHorn, E. & Mahadevan, S., 2021. Digital Twin: Generalization, characterization and implementation. *Decision Support Systems,* 145, pp. 1–11.

Veneri, G. & Capasso, A., 2018. *Hands-On Industrial Internet of Things: Create a powerful Industrial IoT infrastructure using Industry 4.0.* Birmingham, UK: Packt Publishing.

Vermesan, O. et al., 2020. The Next Generation Internet of Things: Hyperconnectivity and Embedded Intelligence at the Edge. In: *Next Generation Internet of Things; Distributed Intelligence at the Edge and Human-Machine Interactions.* Gistrup, DK: River Publishers, pp. 19–102.

Vohra, M., 2023. Overview of Digital Twin. In: *Manisha Vohra.* Newark, NJ: Wiley, pp. 1–18.

Webster, F., 2014. *Theories of the Information Society.* fourth ed. New York, NY: Routledge.

Williams, T. J., 1994. The Purdue enterprise reference architecture. *Computers in Industry,* 24(2–3), pp. 141–158.

Wohlgenannt, I., Simons, A. & Stieglitz, S., 2020. Virtual Reality. *Business & Information Systems Engineering,* 62, p. 455–461.

Wolf, R. & Lepratti, R., 2020. *Smart Digital Manufacturing: A Guide for Digital Transformation with Real Case Studies Across Industries*. Hoboken, NJ: Wiley.

Yildirim, C., 2019. Don't make me sick: investigating the incidence of cybersickness in commercial virtual reality headsets. *Virtual Reality*, 24, p. 231–239.

Zidek, K. et al., 2020. Digital Twin of experimental smart manufacturing assembly system for industry 4.0 concept. *Sustainability*, 12(9), pp. 3658–3673.

3 Gamification types for industrial work

Gamification and industrial production

In the 21st Century, there has been and still is a growing emphasis on gamification, particularly in education and learning, health and fitness programs, and workplace engagement, achieved through the creation of incentive structures via apps or IT systems (Majuri, et al., 2018; Johnson, et al., 2016; Ferreira, et al., 2017). Additionally, it was suggested that industries could utilize gamification strategies (Burke, 2014; Kapp, 2012; Stewart, et al., 2013). Accordingly, Miriam A. Cherry wrote, 'gamification can be used at existing jobs in order to increase worker engagement, especially if those jobs are tedious or repetitive' (Cherry, 2012, p. 53). Worker engagement was just one way to exploit gamification at work, according to Cherry, the others were social networking gamification to get attention and get more users, work that does feel more like fun than work, and games turned into work in e.g. MMOG sweatshops (Cherry, 2012).

Marc Prensky advocated at the beginning of the Century that the next generations would get better business training through digital game-based learning (Prensky, 2001), and although she never mentions the term gamification, Jane McGonigal puts forth the idea that games should not only be used for amusement but also embark on global-scale issues such as climate change, poverty, and global health (McGonigal, 2011). Within this framework, Ole Goethe stated that gamification would work for any given industry (Goethe, 2019). Ole Goethe wrote: 'Tens of millions of users use gamification every day to improve their lives, workplaces, and bottom lines. Because of its unprecedented efficacy, gamification has embarked upon a profitable high-growth vertical, rapidly adding jobs, expanding itself

DOI: 10.4324/9781003406822-3

internationally, and changing the world for better' (Goethe, 2019, p. 24).

Accordingly, there are many expectations for gamification as regards improving industrial work, especially concerning Industry 4.0. Five different perspectives for usage of gamification in Industry 4.0 can be recognized (Reis, Junior og Gewehr). First, gamification may be used to enhance motivation among workers, increasing efficiency, workflow, productivity, and work satisfaction, known as work-joy, which is the happiness of working. Second, gamification may be used to enable immersion in virtual reality and introduce products to clients, raising awareness about new technologies and product lines. Third, gamification may be used for extensive training and boosting human skills. Fourth, developing gamification to enhance servitization, transforming the production into an additional service in continuation of the production.

Emanuelle Savignac has focused more specifically on how games and gamification could enhance the workplace experience, turning work tasks into challenges and using simulation games at the workplace through real-life scenarios and role-playing games during training (Savignac, 2016). That said, the potential transition for gamification in an industrial setting has, until recently, not been fully realized (Garbaya & Lim, 2019). Nevertheless, analysis of Industry 4.0 has shown gamification has proven to be a powerful tool in this new industrial age (Reis, et al., 2020). Still, until recently, it was considered surprising that gamification had not yet expanded into manufacturing and industrial production since a lot of processes have measured outcomes that are typically transferred to business intelligent systems such as enterprise resource planning (ERP) and production planning systems (PPS) and accordingly this could in principle function as the groundwork for gamification applications (Korn, et al., 2015).

Kristi Larson has reviewed serious games and gamification in the corporate training environment, acknowledging that the use of gamification and serious games has intensified over the years (Larson, 2020). In here, Larson also points out that many companies are hesitant to use gamification in the learning system processes. One reason is that any such change requires change management to succeed. Consequently, it is not enough to introduce gamification in the workplace; there must be a well-thought-out plan to roll out the gamification program (see Chapter 4).

Makenzie Keepers, Isabelle Nesbit, David Romero, and Thorsten Wuest examined the current research surrounding gamification for manufacturing (Keepers, et al., 2022). In their study, they confirmed that the field is growing and that the focus has been on ground-floor studies and conceptual design work, pointing out that this groundwork will become a solid base for future research into gamification and the implementation of these applications. Also, they perceived the benefits of gamification for manufacturing as both psychological- and production-oriented. However, they also acknowledge that gamification in manufacturing is not always an easy task. There are limitations as regards finding relevant adoptions of the technology and knowing when to use gamification in the production in Industry 4.0. Besides, they hint at ethical considerations surrounding the use of gamification although this has not been fully developed yet.

In addition, they acknowledge that smart manufacturing and Industry 4.0 play an important role and that an increasing number of technologies are being used at the workplace, which opens the many potential uses of gamification with computers, smartphones, tablets, augmented reality, and virtual reality, while at the same time automation has the opposite effect (Keepers, et al., 2022). Gamification does not make any sense if there are no people involved in the work processes. One could then assume that gamification would go away with gradually more use of automation, however, automation works best with larger lot sizes and consequently smaller companies are not ready to automate production. Conversely, due to increased customization with more variants because of new technological possibilities, even larger companies are going for smaller lot sizes, making automation harder to install and implement (Korn, et al., 2015).

Workplace gamification

In contemporary organizational contexts, the application of gamification has emerged as a significant technique to elevate employee engagement and efficiency. The embedding of game mechanics into fundamental areas such as product and process engineering, production planning, and supply chain design has the potential to convert prosaic tasks into engaging activities, thereby enhancing employee motivation and performance. The juxtaposition of work and play within this framework cultivates an environment where regulatory adherence and productivity are not merely obligatory but are seamlessly woven into the organizational fabric, engendering a more dynamic and resilient workforce.

Workplace gamification on the work floor may be applied to five key areas: 1) product and process engineering, 2) production planning, 3) production, execution, and control, 4) supply chain design and planning, and 5) transportation planning and execution (Warmelink, et al., 2020). All this is connected to an information system, and it is precisely, to a large extent, an information system that is used for gamification in Industry 4.0. They do this because it is easy to obtain data and thus monitor progress. Information systems also offer an interface where gamification may be implemented. This means that we need to understand how an information system works in a workplace, where the systems initially had a practical purpose, unlike, for example, computer games which have primarily been about playing for pleasure. Information systems have traditionally been understood as either utilitarian or hedonic (Köse, et al., 2019). An example of a useful information system could be a SCADA system, monitoring and controlling industrial processes and collecting and analyzing real-time data to optimize operations. On the other hand, a video game would be considered a hedonic information system, allowing players to engage in a virtual world, often featuring challenges, objectives, and narratives (Moore, 2011). When gamification is involved, the line between utilitarian and hedonic information systems becomes blurred. These systems are known as dual information systems in the sense that they are both utilitarian and hedonic. A common way to create a dual information system is to gamify an already existing utilitarian information system (Köse, et al., 2019).

If we follow the idea that information systems become a mixture of utilitarian and hedonic information systems, then we owe it to ourselves to find a way to approach the systems in such a manner that it makes sense to read gamification. Karl M. Kapp suggests that industrial compliance regulations may be used to create gamification because the regulations are rule-based, and these rules may be turned into a game, teaching workers about safety and government regulation, and even keeping the organization operational (Kapp, 2012). That is a way to do it, but it is far from the only one. But the point is good enough that it's about understanding the rules of the information systems because rules are what information systems have in common with games. This means that an entry point to understanding gamification in relation to the factory's information systems is to find similarities with the rules of production in relation to how a game works. This integration can be carried out in different locations and ways within the organization. The use of information systems, initially designed for utilitarian purposes like SCADA systems, now also incorporates hedonic elements, creating dual information systems that blend functionality with enjoyment.

One way to achieve this combination of utilitarian and hedonic work practices is by being subjected to flow as an optimal experience (Csikszentmihalyi, 1991). Tagging along this line of thought, it then follows that if employees are to achieve flow, they need to experience that they can learn the tasks at an easy level in the beginning, but as they become more skilled at their tasks, the level of difficulty should increase correspondingly. Unfortunately, this is not usually the case in factories. There is often a steep learning curve in the beginning, especially for novices, and as it becomes routine for experienced workers, there are sometimes not many opportunities to raise the level of difficulty. Therefore, the work can appear anxiety-inducing at first and monotonous and even boring as the workers develop their skills. These problems may be solved through the use of serious games and gamification.

There have also been considerations about the impact of games in the workplace and gamification on different generations, with the assumption that especially the younger generations would respond well to game-like initiatives (Prensky, 2001). In a best practice study (Caserman, et al., 2024), researchers investigated the generational differences in the acceptance of gamification. The study examined three generations: Generation Z (born between 1995–2009), Generation Y aka. Millennials (born between 1980–1994), and Generation X (born between 1965–1979). The hypothesis was that Generation Z would be more positive towards gamification than Generation Y, who would, in turn, be more positive than Generation X. This was confirmed, as Generation Z was more open to using a gamified app, mini-games, and gaming in a business context, and as predicted, Generation X was the most skeptical about introducing gamification. However, there was a surprise: Generation Y was the most positive towards badges. In the experiment, avatars were used as a gamification element, and everyone found them to be the most positive element. On the other hand, they were neutral towards leaderboards or points. Regarding rewards, they preferred vouchers, trips, or free coffee over virtual rewards. It may be added that Generation Y strives for autonomy and empowerment at the workplace (Schönbohm & Urban, 2014; Goethe, 2019).

Types of gamification and serious games

Gamified experience for industrial production can be categorized into three distinct types, each serving a unique purpose. Instructional work-related games (IWRG) are developed to take employees out of their usual work environments to train them in skills and areas that should ideally be directly related to their work tasks, thereby creating a conducive work environment. Idle time games (ITG), on the other hand, aim to utilize employee's downtime productively by offering casual learning games that can be used to upgrade the workforce. Learning while working (LWW) includes game elements in actual work processes, intending to help employees manage their tasks better, acquire new skills, and hopefully encourage them to complete their tasks faster and more efficiently. This should happen seamlessly as they perform their regular work tasks, allowing for personal development in the workplace. Each of these three gaming experiences supports professional job training in its own way.

Serious games draw inspiration from the video and computer game industry, where game genres such as puzzles, quizzes, simulation games, role-playing games, and strategy games are reimagined in an educational context (Prensky, 2001; Kalmpourtzis, 2018). The objective is to understand how the game genre works and then envision and apply the learning potential to create an educational game based on how the game functions. Gamification, of course, also draws inspiration from the computer game industry, but here only elements from the game are used (Kapp, 2012). Thus, it is more about extracting a selection of game design patterns like game mechanics and other features and finding them useful in a work context (Björk & Holopainen, 2004; Adams & Dormans, 2012; Triantafyllou & Georgiadis, 2022).

Instructional work-related games (IWRG)

Instructional work-related games (IWRG) belong to a category of serious games crafted with a primary focus on educational and training purposes within the context of industrial work and manufacturing (Abt, 1987; Kapp, 2012). The primary objective of IWRGs is to extract employees from their standard production environment, thereby enabling them to engage in simulated work activities or investigate various work-related themes within a controlled and risk-free environment. The central point of these games is that the workers are taken out of the production, and for a while, they are simulating work processes or other kinds of work-related themes.

IWRG can have different purposes. The most common in serious games is skill enhancement and training, giving workers the opportunity to upgrade and boost their knowledge (Abt, 1987). Still, there may be other less obvious purposes like team building and collaboration, innovation and creativity, process improvement and orientation, as well as safety and compliance. These games can manifest in various forms, such as traditional board games, role-playing games, or computer-based simulations. What sets IWRG apart is the core objective of temporarily removing workers from their usual work production environment, allowing them to engage in simulated work processes pertinent to their jobs, or exploring different work-related themes in a controlled and risk-free setting.

A typical instructional work-related game is about finding some situations, work practices, or conflicts in the workplace that can be simulated in some form of game to learn from the process (Wilkinson, 2016). Usually, there will be an introduction to the game and what is going to happen. Next, there are one or more game sessions where participants test and improve their skills or otherwise experience some game-based simulations. This is then followed by a debriefing. The idea is that these instructional work-related games transfer knowledge, skills, and competencies so they can be used in daily life.

Idle time games (ITG)

Idle time games (ITG) are games that can be played during idle time at the factory, contrary to cycle time when workers are actively employed in the production rather than observing or preparing production cycles (Prakash & Aneesh, 2015). Idle time games should not be confused with the video game genre idle games, also known as incremental games, which are video games that require minimal interaction of the player, and yield rewards over time when the player is not actively playing. Idle time games are serious games or gamification made for factory workers that can play these games during idle times while e.g. monitoring work processes.

ITGs are installed in the workplace which means the employees do not have to leave the workplace to play them (Miljanovic & Bradbury, 2023). These games take advantage of Industry 4.0 and the fact that in many cases shop floor workers are monitoring production processes in case anything unfortunate should happen. This means there is idle time that can be used as skill training and knowledge acquisition. These games may be anything the workers can play on their own whether they are puzzles or strategy games. It is, however, important to create a game that you may leave at any given moment. If the games become too immersive, they may get in the way of the production unless, of course, the immersive experience is part of the production. However, in such cases, the games may probably be considered LWW.

In my action research, I have designed some ITGs. The first game was a simple puzzle, which I designed with a few variations. In many factories, assembling and disassembling equipment is necessary. This is done either because the processes require that all equipment be cleaned or because the equipment occasionally needs inspection and/or repair. For new employees, practicing assembly during their breaks can be good exercise. This idea can also be extended to proper attire in a dress-up game. Some factories require workers to wear specific gear and work clothes in different work zones or during various tasks. Here, workers can practice dressing up a game character in the appropriate outfit for different tasks, ensuring they are ready to wear the correct attire as required.

Another example is a complex puzzle in which the employees learn job instructions (JI) and standard operating procedures (SOP). This is called a process game.

A process game is a type of ITG. In its simplest form, it involves developers breaking down a work task into individual components and capturing a picture and description of each process. In many cases, this information can be taken directly from job instructions (JI) and standard operating procedures (SOP). If there are no pictures for all procedures, it will be necessary to take more pictures so that each sub-procedure is illustrated or photographed. These can be either still images of the process or short animated film sequences. Then, all the pictures and job instructions are put into a sequential series where text or possibly audio explains each image's subtask.

At this point, it should be possible to remove parts of the series and place the pictures in the incorrect order. It is then up to the employee, using the images and text, to determine the correct order by moving the pictures around with the mouse.

Initially, the employee who is learning the work process only needs to arrange 5–7 subtasks in the correct order. Once they can do this, the difficulty increases with more subtasks where the employee can read the text and look at the picture to determine the order. Gradually, there will be more and more pictures with subtasks that the employee must consider and put in the correct order. Ultimately, the task is to arrange all the pictures in the correct sequence without help from text or audio. When the employee can do this and has thus learned the entire work process, it becomes routine to know the sequence of the standard operating procedure.

Process games are particularly suited for novice employees who need to learn new work processes or experienced workers who need to refresh standard operating procedures that they have not performed for a long time or only perform infrequently. This way, employees can use their breaks or idle time to learn or remember the procedures' sequence.

Learning while working (LWW)

Learning While Working (LWW) means that employees learn while they work. This implies that they learn naturally through work processes and the reward systems embedded through gamification (Jacob, et al., 2022). LWW serves as an alternative to sending employees to courses or peer training, as gamification of work processes in itself creates immediate learning. Either this can occur through superficial BLAP, or it can involve a fundamental reconstruction of work processes, as seen in deep gamification.

It can be relatively easy to add superficial BLAP gamification on top of existing systems as an extra layer that does not disrupt the actual work processes and cyber-physical systems (Nicholson, 2015). However, if we want to take a more ambitious approach and change work processes and routines to transform them into gamified learning, it will require a thorough reorganization of the entire workflow on the factory floor. This will necessitate extensive research and development to ensure it works as intended. To transform work processes into gamified learning, it is imperative to undertake a comprehensive reorganization of the entire workflow on the factory floor, and it must be noted that there is no assurance of its successful implementation.

The question then is whether gamification should change the SCADA system, which in many cases will require significant effort or might not be feasible at all. In most cases, there is no chance that designers can deliver changes to the SCADA system. Alternatively, it may be sufficient to create a layer on top of the SCADA system where employees can see their progress or create a simulation or digital twin with live data from the production or the historian database. Unless, of course, the SCADA system becomes gamified from the beginning. One could assume that maybe future SCADA systems will incorporate integrated features designed to facilitate the creation of gamified experiences.

LWW requires that these progress points can be evaluated to give points or otherwise assess performance. In traditional computer game design, the computer calculates whether a given action should be rewarded or possibly punished based on given algorithms. The computer thus acts as a judge to evaluate actions performed in the game. If a computer on a factory floor could make such an evaluation, then the work process could, in principle, be automated. Therefore, it is not feasible to leave this evaluation to the computer system because as soon as a process is automated, gamification no longer makes sense since humans are no longer involved in the process. Consequently, the only ones who can assess the correctness and value of a given action will be other employees.

The computer system can handle the assignment of points, but other employees must evaluate whether the action was performed correctly. However, there is a way for the system to assess a process. If an action results in something working for a long time without breaking down, this can be recorded in the system. If it turns out to work over a long period, the employee can be awarded points. The downside of this method is that feedback is so delayed that the employee may have difficulty relating to it when it finally comes.

Pit stop games

A pit stop game is a form of gamification where the goal is to get everything done within a specific timeframe – similar to a racecar making a pit stop during a race. A racecar coming into the pitstop gets tires changed in 2–3 seconds. This is well-executed efficiently and with high precision. A pit stop game promotes the same kind of efficient work mentality. It emphasizes accuracy and teamwork.

Therefore, three factors are crucial for success in a pitstop game: teamwork, proper timing, and high quality. This type of game can be used for LWW.

Timing can, of course, be measured by time, but it is not enough if only one person is fast and the rest lag behind because they cannot keep up. It is important to evaluate it as a collective performance where everyone completes his or her tasks on time.

Quality, unfortunately, can be more challenging to assess and may require historical data. For instance, if an item has been repaired and takes a long time before it needs another repair, it is a success. However, if it breaks down again quickly, it is a failure.

This means, regrettably, that feedback is delayed until data can determine whether it was a success or failure. The first information received will be if a mistake was made. When it takes a long time before the worker receives feedback, the impact is unfortunately diminished. This might sometimes be the only viable solution, but it can mean that the effects of gamification are absent.

A variant of the pit stop game is only looking at time because this factor is the easiest to measure. However, this may lead to perverse incentives due to a lack of focus on quality. Leading to poor-quality products. Another variant is to make it into an individual game rather than as a team effort. This makes sense if the work being done is not based on teamwork.

Dynamic Gantt planning game

A Gantt chart is a for scheduling work tasks, providing a visual overview of the tasks and showing how they might be interdependent (Kumar, 2005). Some tasks can only be performed if certain other tasks are completed. Within project management, a dynamic Gantt chart acts as an interactive tool, rendering a visual depiction of tasks, timelines, and dependencies, which allows for real-time updates and adjustments (Renna, 2013; Yin, 2016). Instead of employees physically adjusting the Gantt chart and manually linking different tasks, this is done automatically in a dynamic Gantt. Because it is computer-controlled, it offers unique features that the traditional Gantt chart did not have. The dynamic Gantt is interactive, receives automatic updates from connected information technology and operational technology (IT/OT) systems, and everything happens in real-time. Hence, the dynamic Gantt chart is capable of managing both task planning and resource allocation in real-time, making it notably beneficial for project management, resource distribution, and scheduling.

From a gamification perspective, the dynamic Gantt provides unique opportunities to monitor workplace activities and allows for the incorporation of game elements that promote beneficial behavior for optimizing production. In this sense, the dynamic Gantt is a real-time overview of upcoming, current, and completed tasks. With such a tool, it becomes possible to get an overview of all tasks and their sequence. This way, the game developer can understand the entire factory as procedures and rules in a game, creating the basis for attaching reward systems to promote industrial efficiency in a dynamic Gantt planning game.

Following the gist of LWW, a dynamic Gantt planning game can be used as a way to gather information around the clock about work processes at the factory. With this information, it is possible to put up an extra layer as a digital twin or simulation, adding the necessary game elements to the human-machine interaction. This means that the workers, in principle, can get instant feedback from the system based on their decisions. Alternatively, they may at least obtain deferred feedback predicated on data from the historian database.

Advantages and disadvantages of serious games and gamification

The obvious advantage of IWRG is, of course, to enhance certain skills and provide the training needed to do specific tasks. Designing a safeguarded setting where actions and mistakes are allowed without dire repercussions, and free from the daily grind, encourages skill enhancement via practical techniques; while progress is valuable, this strategy notably enhances employees' problem-solving skills and clarifies workplace dynamics, stressing that to achieve success, one must value progress and recognize this technique's capacity to improve problem-solving and workplace understanding. Additionally, the use of IWRGs may offer more entertaining while retaining learning experiences compared to conventional methods. If this edutainment works, because the employees find these games motivating and memorable, then it can lead to better retention of the acquired skills and knowledge. If the games in question have cooperative elements, there will also be the possibility of learning collaboration, which can translate to improved teamwork.

Nevertheless, IWRGs are not without disadvantages. These games can be resource-intensive because they allocate a lot of time and workforce. Taking employees out of their regular undertakings for game sessions may lead to a temporary decrease in productivity. The intended performance of games also involves a great deal of time and effort to ensure they work as planned. It is not enough to simply come up with the idea; continuous refinement and improvement are necessary, which, as mentioned, takes time.

The ITG process game also has the advantage that developers can set a difficulty level, where it is very easy at the beginning, and over time becomes more difficult by increasing the complexity according to the individual's skills. This means that employees can be challenged at the level they have reached, thereby supporting the zone of proximal development from a pedagogical standpoint and finding the right level of flow from a game design perspective. Experienced users who just need to refresh their skills do not need to start the game from scratch but can choose an appropriate difficulty level themselves.

However, the ITG game can also be used to check if experienced employees follow the job instructions or standard operating procedures. In some cases, it may not matter the exact order in which things are done as long as the end result is correct, but in other cases, it is crucial that the process sequence is followed exactly. The fact is that bad habits can be passed down through on-the-job training, and here a process game can help ensure that everyone learns the same procedures in exactly the same order.

An advantage of both puzzle games and process games is that they are fairly easy to maintain and update. Over time, new items will replace older items at the workplace. Likewise, standard operating procedures and job instructions may be changed due to technological changes, new regulations, and optimization. By changing the pictures and descriptions, it can all be put into the same game system. It is important that these games can be maintained effortlessly; otherwise these games will only be used for a short time and then be forgotten.

The biggest immediate advantage of LWW is that workers can stay on the factory floor and learn during the process. However, this is an advantage that LWW shares with idle time games (ITG). The difference is that LWW does not utilize breaks and idle time for learning. Thus, it becomes more efficient in principle because one actually learns while working.

The disadvantage of LWW is that it requires a significant amount of effort to make it work, and it must function in real-time without issues. Developers also need access to sensitive data, where it is crucial to ensure that it does not fall into the wrong hands. Additionally, it can lead to increased monitoring of workers' performance. Furthermore, it is important to ensure that LWW does not interfere with the actual work that needs to be done.

ITG and especially LWW must work within the information system and even the cyber-physical system of the factory. This means that these activities may require access to data from sensors, SCADA, and human-machine interfaces. It is important to get a full notion of what kinds of access are required for the specific gamification product.

Purdue Enterprise Reference Architecture (PERA) in relation to gamification

Purdue Enterprise Reference Architecture (PERA) details the necessary data and information levels for gamification tailored to shop floor workers, middle management, or executives (Schekkerman, 2004; Xu, et al., 2023). Since gamification can be utilized at all levels of the organization, it is important to consider the unique challenges and opportunities that each level within the company encounters (Gudiksen & Inlove, 2018). To do this, it will be necessary to address each level of the PERA model (see Chapter 2).

At level 0, it does not make much sense to talk about gamification as it primarily concerns machines. However, it should be added that there are people who service the machines, and it can be considered that their actions might be the basis for gamification, but this will first be registered at level 1. There is also the possibility of creating games or game-like features using analog games (Abt, 1987; Grace, 2020; Mazarakis, 2021).

As we ascend to level 1, sensors capture real-time data from the production floor, forming the basis for creating LWW gamification. This level is crucial as it marks the initial point where gamification can be effectively applied to enhance learning and engagement on the shop floor (Hellebrandt, et al., 2018; Tsourma, et al., 2019; Sochor, et al., 2021; Dolly, et al., 2024).

At level 2, the integration of human-machine interfaces (HMI) allows employees to monitor and control production processes. LWW gamification should focus on these employees and how they are able to handle production. This is where we find SCADA, which can be subject to optimizing human use of the system, either by introducing gamification in SCADA or by adding extra layers through simulation and digital twins (Bucchiarone, 2022; Ulmer, et al., 2022; Ismagilova, et al., 2023). It is also possible to create ITG for this level of human involvement in production.

At levels 3 and 4, the focus shifts to planning and management through manufacturing execution systems (MES) and enterprise resource planning (ERP) systems, respectively. Both types of systems can be subject to the gamification of their tasks with LWW, in which the gamification application promotes motivation and optimizes human behavior. Learning how to use the systems could be done using either IWRG or ITG.

The PERA model serves as a foundation for incorporating gamification within Industry 4.0. However, it is important to note that Industry 4.0 has posed challenges to the model. This is not because the model is rendered useless but because new opportunities, such as drones and tablets, have emerged and must be integrated into the model or rethought to maintain a high level of security (Xu, et al., 2023). This leads to an exploration of how gamification leverages the new opportunities presented by Industry 4.0.

Industry 4.0 and gamification

When we look at the possibilities for gamification and the utilization of Industry 4.0, it becomes clear that the three different forms of gamification exploit the potential differently. Generally, learning while working (LWW) will be most dependent on well-executed Industry 4.0 applications and cyber-physical systems (CPS). On the other side of the range, instructional work-related games (IWRG) are typically less dependent on Industry 4.0 and in many cases; they do not need Industry 4.0 at all. However, it should be noted that it depends on the specific setup. Hence, if an LWW only utilizes Industry 4.0 to a lesser degree, while an IWRG is created in such a way that it is completely dependent on Industry 4.0, it does not follow the general trend.

If a typical SCADA is to be utilized for gamification, it will usually be where humans interact with the system, which is through the human-machine interface. It is simply that it can be simulated so that it can be presented through an IWRG. The same applies to an ITG, but here it will be much more possible to exploit live data, as they do not leave the workplace, which creates greater security.

LWW will, in many cases, be directly dependent on live data from the processes and, at a minimum, get them from the historian database. The extent to which a typical gamification application can have direct contact with cyber-physical systems is limited. It will generally be through a human-machine interface (HMI) as with SCADA in general. On the other hand, the Industrial Internet of Things (IIOT) offers some possibilities. Here, smartphones and tablets can be connected, making it possible to insert a game like ITG or add game elements, so the use of these tablets and smartphones gets an extra layer where the worker can be rewarded for the desired behavior.

As for big data and artificial intelligence, it may seem cumbersome to apply, and this is also where we see many creative developments in these years. However, it should be noted that if a process can be automated based on big data and artificial intelligence, there would be no need for humans, which means that it does not make sense to implement gamification. However, it might still be possible to use big data to consider users' behavior in gamification, which can form the basis for automation. Therefore, in this way, gamification, where behavior in the system becomes data, can itself feed artificial intelligence for the automation of work tasks. Artificial intelligence can also be used in game design to create bots that have natural language and possibly create relevant images and music for the design.

Augmented reality can be used with smartphones and tablets to add an extra layer over reality and this layer can be games or game-like applications. It will be especially useful in LWW games, where gamification can be added in real-time. Nevertheless, it will also be applicable to ITG, where the extra layer of reality can make the games and the connection realistic, but it must be done correctly. Regarding virtual reality, it will be better suited for IWRG, where it is possible to completely immerse oneself in the virtual reality and learn through that experience, but in most cases, it will not be functional in a work process to completely immerse oneself in a virtual world, at least not at the stage where technology is now. It will lack the necessary input from the surrounding environment. It's not so much that it cannot be done technologically, but a person in virtual reality becomes too isolated from practical everyday life.

Cloud computing is an option to store data so that it can be retrieved from many different positions, both from stationary and mobile devices. All that data can naturally be used as a basis for gamification. The more data a game developer has access to, the more possibilities there are to develop games and game-like applications.

A simulation or digital twin is an excellent tool for creating a safe space where workers can juggle live data without any errors affecting the actual production. Here, it is possible to experiment with how production can be optimized. It is a beneficial tool because live data gives a sense of presence, immediacy, and urgency, which helps provide a realistic experience. Live simulations and digital twins will be particularly useful for ITG and LWW, where it will be possible to use live data while production is running, and one is present in the factory. This creates an opportunity to conduct very realistic scenarios.

Summary

It is possible to enhance engagement, motivation, and productivity through the intelligent use of gamification in industrial production, especially within network-based and digitalized smart manufacturing. Indeed, it will be a way to take advantage of the many new technologies associated with Industry 4.0. This is particularly evident in the PERA model, where gamification can be introduced, and serious games can be applied at multiple levels.

There are three types of integration of gamification in manufacturing companies. IWRG can be used for serious games aimed at skill acquisition. ITG can be serious games played in factories during downtime to gain knowledge. LWW is true gamification, where the actual work processes are monitored, and there is an opportunity to provide feedback, preferably immediate feedback.

With this overview of how gamification and serious games can work in smart manufacturing and Industry 4.0, the next step is to see how gamification and serious games can be designed for the manufacturing industry.

Bibliography

Abt, C. C., 1987. *Serious Games.* Lanham, MD: University Press of America.

Adams, E. & Dormans, J., 2012. *Game Mechanics: Advanced Game Design.* Berkeley, CA: New Riders Games.

Björk, S. & Holopainen, J., 2004. *Patterns in Game Design.* Hingham, MA: Charles River Media.

Bucchiarone, A., 2022. Gamification and virtual reality for digital twin learning and training: architecture and challenges. *Virtual Reality & Intelligent Hardware*, 4(6), pp. 471–486.

Burke, B., 2014. *Gamify: How Gamification Motivates People to Do Extraordinary Things.* Brookline, MA: Bibliomotion.

Caserman, P., Baumgartner, K. A., Göbel, S. & Korn, O., 2024. A best practice for gamification in large companies: An extensive study focusing inter-generational acceptance. *Multimedia Tools and Application*, 83, pp. 35175–35195.

Cherry, M. A., 2012. The gamification of work. *Hofstra Law Review*, 40(4), pp. 851–858.

Csikszentmihalyi, M., 1991. *Flow: The Psychology of Optimal Experience.* New York, NY: Harper Perennial.

Dolly, M., Nimbarte, A. & Wuest, T., 2024. The effects of gamification for manufacturing (GfM) on workers and production in industrial assembly. *Robotics and Computer-Integrated Manufacturing*, 88, pp. 1–20.

Ferreira, A. T., Araújo, A. M., Fernandes, I. & Miguel, I. C., 2017. *Gamification in the Workplace: A Systematic Literature Review.* Cham, CH: Springer, pp. 283–292.

Garbaya, S. & Lim, T., 2019. Introduction to the special issue on "Gamification of Industrial Systems". *International Journal of Serious Games*, 6(2), pp. 21–22.

Goethe, O., 2019. *Gamification Mindset.* Cham, CH: Springer.

Grace, L. D., 2020. *Doing Things with Games: Social Impact through Play.* Boca Raton, FL: CRC Press.

Gudiksen, S. & Inlove, J., 2018. *Gamification for Business: Why Innovators and Changemakers use Games to Break Down Silos, Drive Engagement and Build Trust.* London, UK: Kogan Page Publishers.

Hellebrandt, T., Ruessmann, M., Heine, I. & Schmitt, R. H., 2018. *Conceptual Approach to Integrated Human-Centered Performance Management on the Shop Floor.* In: *Advances in Human Factors, Business Management and Society: Proceedings of the AHFE 2018 International Conference on Human Factors, Business Management and Society, July 21-25, 2018, Loews Sapphire Falls Resort at Universal Studios (vol. 9). Orlando, Florida, USA:* Springer International Publishing, pp. 309–321.

Ismagilova, G., Lysenko, E. & Bozheskov, A., 2023. *Gamification in Industry: Simulation-Game Modeling of Production Processes.* Cham, CH: Springer.

Jacob, A., Faatz, A., Knüppe, L. & Teuteberg, F., 2022. Understanding the effectiveness of gamification in an industrial work process: An experimental approach. *Business Process Management Journal*, 28(3), pp. 784–806.

Johnson, D. et al., 2016. Gamification for health and wellbeing: A systematic review of the literature. *Internet Interventions*, November, 6, pp. 89–106.

Kalmpourtzis, G., 2018. *Educational Game Design Fundamentals: A Journey to Creating Intrinsically Motivating Learning Experiences.* Boca Raton, FL: CRC Press.

Kapp, K. M., 2012. *The Gamification of Learning and Instruction: Game-Based Methods and Strategies for Training and Education.* San Fransisco, CA: Wiley.

Keepers, M., Nesbit, I., Romero, D. & Wuest, T., 2022. Current state of research & outlook of gamification for manufacturing. *Journal of Manufacturing Systems*, 64, pp. 303–315.

Korn, O., Funk, M. & Schmidt, A., 2015. Towards a gamification of industrial production: A comparative study in sheltered work environments. In: *EICS '15: Proceedings of the 7th ACM SIGCHI Symposium on Engineering Interactive Computing Systems.* New York, NY: Association for Computing Machinery, pp. 84–93.

Köse, D. B., Morschheuser, B. & Hamari, J., 2019. Is it a tool or a toy? How user's conception of a system's purpose affects their experience and use. *International Journal of Information Management*, 49, pp. 461–474.

Kumar, P. P., 2005. Effective use of Gantt chart for managing large scale projects. *Cost Engineering*, 47(7), pp. 14–21.

Larson, K., 2020. Serious games and gamification in the corporate training environment: A literature review. *Tech Trends*, 64(2), pp. 319–328.

Majuri, J., Kovisto, J. & Hamari, J., 2018. Gamification of education and learning: A review of empirical literature. In: *Proceedings of the 2nd International GamiFIN Conference.* Pori, Finland: CEUR, pp. 11–19.

Mazarakis, A., 2021. Gamifcation reloaded: Current and future trends in gamification science. *I-com*, 20(3), pp. 279–294.

McGonigal, J., 2011. *Reality Is Broken: Why Games Make Us Better and How They Can Change the World.* New York, NY: Penguin Books.

Miljanovic, M. A. & Bradbury, J. S., 2023. Engineering Adaptive Serious Games Using Machine Learning. In: *Software Engineering for Games in Serious Contexts: Theories, Methods, Tools, and Experiences.* Cham, CH: Springer Nature, pp. 117–135.

Moore, M. E., 2011. *Basics of Game Design.* Boca Raton, FL: CRC Press.

Nicholson, S., 2015. A RECIPE for meaningful gamification. In: *Gamification in Education and Business.* Cham, CH: Spring, pp. 1–20.

Prakash, A. J. & Aneesh, K. S., 2015. Cycle time and idle time reduction in an engine assembly line. *International Journal of Science Technology & Engineering*, 2(5), pp. 139–143.

Prensky, M., 2001. *Digital Game-Based Learning.* New York, NY: McGraw Hill.

Reis, A. C. B., Junior, E. S., Gewehr, B. B. & Torres, M. H., 2020. Prospects for using gamification in Industry 4.0. *Production*, 30, e20190094. https://doi.org/10.1590/0103-6513.20190094.

Renna, P., 2013. *Production and Manufacturing System Management: Coordination Approaches and Multi-Site Planning: Coordination Approaches and Multi-Site Planning.* Hershey, PA: Engineering Science Reference.

Savignac, E., 2016. *The Gamification of Work: THe Use of Games in the Workplace.* Hobroken, NJ: Wiley.

Schekkerman, J., 2004. *How to Survive in the Jungle of Enterprise Architecture Frameworks: Creating Or Choosing an Enterprise Architecture Framework, Second Edition.* Victoria, BC: Trafford Publishing.

Schönbohm, A. & Urban, K., 2014. *Can Gamification Close the Engagement Gap of Generation Y?: A Pilot Study from the Digital Startup Sector in Berlin.* Berlin, DE: Logos Verlag Berlin.

Sochor, R., Schenk, J., Fink, K. & Berger, J., 2021. Gamification in industrial shopfloor – Development of a method for classification and selection of suitable game elements in diverse production and logistics environments. *Procedia CIRP*, 100, pp. 157–162.

Stewart, J. et al., 2013. *The Potential of Digital Games for Empowerment and Social Inclusion of Groups at Risk of Social and Economic Exclusion: Evidence and Opportunity for Policy.* Seville, ES: European Union.

Triantafyllou, S. A. & Georgiadis, C. A., 2022. Gamification design patterns for user engagement. *Informatics in Education – An International Journal*, 21(4), pp. 655–674.

Tsourma, M. et al., 2019. Gamification concepts for leveraging knowledge sharing in Industry 4.0. *International Journal of Serious Games*, 6(2), pp. 75–87.

Ulmer, J., Braun, S., Cheng, C.-T. & Wollert, J., 2022. Usage of digital twins for gamification applications in manufacturing. *Procedia CIRP*, 107, pp. 675–680.

Warmelink, H. et al., 2020. Gamification of production and logistics operations: Status quo and future directions. *Journal of Business Research*, 106, pp. 331–340.

Wilkinson, P., 2016. *A Brief History of Serious Games.* Cham, CH: Springer, pp. 17–41.

Xu, W., Gao, Y., Yang, C. & Chen, H., 2023. *An Improved Purdue Enterprise Reference Architecture to Enhance Cybersecurity.* New York, NY: ACM, pp. 104–109.

Yin, R., 2016. *Theory and Methods of Metallurgical Process Integration.* London, UK: Elsevier.

4 Designing gamification for industrial purposes

Experience design

Gamification and serious games are special cases of experience design, where the focus is on motivation and learning (Routledge, 2016; Goethe, 2019). If we understand this in a broader context, then game elements can be one way to create motivation and learning but not necessarily the only way. There can be narrative and aesthetic forms that designers can employ, and this is important to keep in mind, because the goal is what is vital. The reason that games and game elements have become so significant is that they are fundamentally interactive and participatory, which is easy to introduce into interactive media. That is why games and game elements have become more significant. This should be compared with the traditional media such as radio and television, which have focused more on storytelling and shows. However, for an experience designer, all tricks in the drawer may apply (Gilbert, 2016; Clark, 2021). Nevertheless, games and game elements integrate more naturally with interactive media and that is why it has become so important for experience designers to understand how games work. When designing gamification and serious games, the experience designer is primarily a game designer. However, for gamification of smart factories, the game designer cannot do this alone. There is a need for a team with many different skills.

Development team

When assembling a team to develop gamification for Industry 4.0, it is important to recognize that different skills will be needed: engineers,

DOI: 10.4324/9781003406822-4

software designers, visual and audio designers, and game designers (Visser, et al., 2016; Fullerton, 2018; Tennant, 2022).

Every project needs a project manager or team leader, and it is required that this person has strong communication and leadership skills (Pedersen, 2009; Thompson, 2018). Additionally, it would be beneficial if they have a background that combines engineering knowledge, software design, and game design, enabling them to understand the different ways of approaching the tasks. Furthermore, the project manager must be able to communicate with a variety of people, from company executives and factory workers to the various team members. The person should possess natural authority and be convincing in the role of project manager.

It is also important to have a game designer (Pedersen, 2009; Fullerton, 2018). A game designer typically needs to understand and create exciting games; however, this alone is not sufficient for gamification in the industry. Of course, the game designer must still have a good understanding of game experiences and game dynamics, but additionally, the game designer must understand how a factory operates and how to increase employee motivation and learning.

A functional team should also include an industrial engineer who knows and understands how a factory works and has in-depth knowledge of industrial production. The kind of engineer may vary depending on the type of factory. There should also be a software engineer who understands how to lead a development team. If the right skills are present, an industrial engineer and a software engineer can be the same person.

The project manager, game designer, industrial engineer, and software engineer form a management team that collaborates to create gamification products. They should utilize their different expertise together. The project manager should listen to the team, be able to mediate if necessary, and make decisions when needed. The game designer has the decisive word on how game experiences and game mechanics work; the industrial engineer is responsible for understanding what will work in an industrial manufacturing context, while the software designer is responsible for making the computer program work and informing about potentials and limitations.

In the team, there will be a need for people of many different backgrounds, skills, and competencies (Tyler, 2015; Merholz & Skinner, 2016; Fullerton, 2018). There will be a need for skilled workers with an engineering background, preferably also with programming skills. This could be systems engineers, quality engineers, or engineers tied to the specific form of production. There may also be a need for a field engineer to implement and maintain the system on-site. Of course, there will also be a need for software developers and programmers who can program the game mechanics and the gamification solution. Both back-end and front-end developers will be required. There may also be a need for cloud engineers and cybersecurity specialists, depending on the specific tasks the development team faces. Regarding design, there is a need for technical artists and sound designers as well as graphic designers. It would also be a good idea to have a UX/UI designer and a QA (quality assurance) tester. The list is not exhaustive, and it will naturally depend on the specific type of gamification the team undertakes. But overall, there is a need for engineers, software developers, and people with a design background.

Development models

The waterfall model in industrial design is a sequential model that involves going through a series of phases to ultimately deliver a finished product. The phases typically include 1) requirement gathering, 2) system design, 3) implementation, integration, and testing, 4) deployment, and 5) maintenance (Badiru, 2019; Cadle & Yeates, 2008; Hilburn & Towhidnejad, 2021). There are clear advantages to this model, namely a clear structure and understanding of each phase. It is relatively easy to estimate timelines and budgets, and the documentation of the design is maintained at every step along the way. However, the waterfall model has its weaknesses (Gomaa, 2011; Liu, 2018). It is rigid, making it difficult to accommodate changes within a tight schedule. Testing of the product occurs late in the process, which can make fixing issues cumbersome and potentially expensive. Consequently, users and stakeholders do not have the opportunity to provide feedback until very late in the process. In the worst-case scenario, this can result in a final product that does not meet their needs, leading to a suboptimal or, in the worst circumstances, an unusable product.

Iterative agile design, also known as agile development, is a method where multiple phases are repeated in a cycle (Douglass, 2021). Each completed cycle is called an iteration, and the final development of a product typically undergoes many iterations. The phases in an iterative design usually begin with initial planning. This leads to the first phase in the iteration process, which involves planning and requirement gathering. The second phase consists of analysis and design, as well as implementation. In the third phase, the preliminary product is tested, leading to the fourth phase, where the preliminary product is evaluated. These four phases are repeated until the product is complete and ready for deployment.

The iterative agile design approach is advantageous because it is flexible, offers early and continuous feedback, focuses on the user, and achieves incremental improvements (Cockburn, 2007). In effect, the project is endowed with the capacity for adaptive modification as developers, through experiential learning, enhance their comprehension of what the product ought to embody and accomplish during the developmental continuum. The disadvantages are that it requires discipline and experience to prevent it from falling apart and becoming unmanageable (Meyer, 2014). The product requirements may be unclear, and it requires a lot of communication along the way, which can be disruptive. The greatest uncertainty is that it is unclear when the project will be finally completed.

It is not necessarily the case that iterative agile design is better suited for developing gamification or vice versa. Each has its own advantages and disadvantages.

The waterfall model is suitable to use if the developers have a clear concept, where it is about conducting a preliminary study with requirement specifications, system design, coding, and testing, and then maintaining the system. However, if this is not the case, iterative design can be an alternative. Here it is about continuously developing and testing ideas, and this can be necessary because game design usually requires developers to test whether game mechanics and audiovisual design have the desired effect on the user. It is not only about testing whether the game works functionally, but also whether it creates the right motivation and encourages the user to change behavior. Furthermore, it must be tested whether the game does not create so-called perverse incentives, where users develop unwanted behaviors that the game unintentionally rewards.

This means that the waterfall model is clearly preferred if the developers have a clear and distinct concept that needs to be delivered to the client and implemented in production. On the other hand, if the developers are more exploratory in their approach with a view to investigating the possibilities in the gamification of work processes, an agile iterative method is preferable.

Gamification for the workplace

When designing for gamification in the workplace, the specific context of gamification is, of course, the particular workplace in question (Salzman & Rosenthal, 1994; Savignac, 2017). This means that gamification must either be specifically tailored to the workplace or adapted from a general gamification solution to the unique issues that the workplace faces. Clearly, if similar issues are found across multiple workplaces, it is possible to present a standardized solution that can be produced on a larger scale. However, gamification for highly specific problems must be developed for that particular type of workplace, which is often more expensive.

The goal is to develop gamification that can be integrated into the systems of the smart factory (Reis, et al., 2020; Keepers, et al., 2022). During this development process, it is crucial to bear in mind that gamification should not interfere with work processes but instead supports them and fosters employee development (Ulmer, et al., 2022). The key point is that gamification should motivate employees; however, if implemented incorrectly, it can, unfortunately, have the opposite effect and demotivate them.

One common way to motivate is through rewards (Goethe, 2019). Rewards can take many forms in gamification at the workplace. They can be abstract, like points and badges, and they can give some opportunities or even be gifts. When giving out rewards, no matter what the form, there should be upheld a sense of fairness. Especially if there should be some sort of competition among the workers. If they are not treated equally in terms of commitment and merits, then there will be build negative tension (Leclercq, et al., 2020). Also, the development team should carefully consider whether rewards should be public or private among the workers themselves. Another issue is that a factory can, from time to time, have breakdowns, and if the progress and rewards fail to appear because of external factors or circumstances, then the gamification risks being perceived as unfair. That is why rewards should never be taken lightly.

Another way to motivate employees is through autonomy (Goethe, 2019). To do this it is important to have decision-making. Decision-making should always have the attention of game development because decision-making is central to game design (Costikyan, 2002). To put it another way, if there is no decision-making, there is no agency, and without agency, there is no way the users can influence the gamification application. Fortunately, smart manufacturing has many situations in which decision-making comes into play. This happens through the interlinking of equipment and devices, making decisions on operational efficiency and product quality based on a data-centric strategy from SCADA, AR, or VR. This means that it will be necessary for efficient, astute, real-time decision-making.

Effective communication for gamification in factories

In connection with gamification at a smart factory, it is important to communicate clearly and distinctly with management about what it sees as the problem at the factory, what might be causing it, and it is also crucial for future collaboration that there is an alignment of expectations regarding what can be achieved. Gamification is a tool that, if executed correctly, can solve specific issues related to motivation and learning, but it is not a magic wand that will solve all conceivable problems.

Once an agreement is in place and access to the smart factories and the necessary IT systems has been obtained, the actual work begins. Regardless of whether the developers use the waterfall model or the agile method, it is necessary to investigate how the specific factory is doing and how the issues can be resolved. It will often be some of the same kinds of problems with motivating workers to perform tasks enthusiastically and precisely, or training employees to perform certain tasks. However, each factory has its own particular issues and a unique culture that must be considered (Suhartini, et al., 2024).

Knowledge about how the smart factory operates can be obtained through observation and interviews (Schensul, et al., 1999). Observation and interviews should be focused on obtaining knowledge about how the factory works. However, before starting with that, the developers should read and understand job instructions, standard operating procedures, and timetables to get an understanding of how production takes place. Consequently, it will either be necessary to conduct observation or interviews and sometimes this can be done simultaneously. Observation should be focused. It is about understanding work processes, how they work, and what can go wrong. Pay special attention if something deviates from standard operating procedures or job instructions, or if there are uncertainties about how things are done. This may be because the workers choose to do it their own way, or there is disagreement about what constitutes best practice. It is not so much about how the machinery specifically works and if it fails sometimes, but rather about how the workers perform their tasks, how the interaction functions between the employees, and how the interaction between computer systems and the workers functions.

Regarding interviews, one should clarify their own role, and it is important to understand how the workers communicate (Dollins & Stemmle, 2021). They can sometimes be skeptical about what gamification might mean, and it is important to put oneself in their position and explain what they can gain from it. Communication problems can also arise because the interviewer and the workers speak from different perspectives. An example could be that the interviewer thinks very abstractly and in terms of systems and solutions, while the interviewed worker relates concretely to their tasks and speaks in terms specific to the company culture, they are part of. Here it is crucial that the interviewer is open-minded and tries to understand the worker's perspective on things, making an effort to speak a language of equality without coming across as superior or condescending. The workers are the experts in the daily work processes, and the interviewer must respect that.

Strategic implementation of gamification

It requires a strategy to launch gamification in the workplace (Herzig, et al., 2015). Many things can go wrong along the way, so it should, therefore, be perceived as trivial to roll out gamification. Think of the process as change management. Suddenly, there is a change, and some may be excited about the change, while others see it as a nuisance.

Kurt Lewin suggests a model for change management in which there are three phases (Achterbergh & Vriens, 2019). The first phase is unfreeze, where the manager prepare employees for the change. The second phase, the change, is the period where the change takes place, and the third phase is refreeze when the change has taken place, and it stabilizes and becomes the new norm. This theory is from the 1950s when there weren't so many changes in workplaces. But we live in a time when changes have become more frequent, which is why other more complex models have emerged.

Jeff Hiatt and Tim Creasey have therefore developed a model with five parts called the ADKAR model, which consists of (1) awareness, where it becomes clear why there is a need for change (Hiatt & Creasey, 2003), (2) desire, where the employees' desire for change is created and fostered (3) knowledge, where employees get the necessary knowledge to handle and work within the framework of the upcoming change, (4) ability, where it becomes possible to implement the change, and, finally, (5) reinforcement, where efforts are made to ensure that changes are reinforced and maintained. The purpose of this model is to get employees on board with changes and get them to carry the change through.

While Kurt Lewin's model was too simplistic and the ADKAR model focused on the individual's approach to change, John Kotter developed an 8-step model that took the starting point in the organization and the creation of a new organizational culture through the change process. Here are the eight steps (Applebaum, et al., 2012; Kotter, 2012):

1. Establish a sense of urgency about the need to achieve change. Individuals are unlikely to undergo transformation if they do not perceive a requisite imperative for such an alteration.
2. Create a guiding coalition. Form a consortium comprising individuals possessing power, dynamism, and influence within the organization to spearhead the change initiative.
3. Develop a vision and strategy. This initiative aims to explicate the essence of the modification, justify the imperative for its implementation, and outline the procedural framework for its execution.
4. Communicate the change vision. Communicate the rationale, specifics, and methodology of the changes at every conceivable opportunity and through every possible medium.
5. Empower broad-based action. Incorporate participants into the change effort, motivating them to reflect on the changes and how to accomplish them, rather than on their aversions and methods to obstruct them.
6. Generate short-term wins. Observing the ongoing transformations and acknowledging the efforts exerted by individuals in effectuating these changes is of paramount importance.
7. Consolidate gains and produce more change. Foster momentum for progress by capitalizing on successful change initiatives, energize

individuals through the transitions, and develop individuals into catalysts for change.
8. Anchor new approaches in the corporate culture. This is fundamental to the prolonged success and entrenchment of the changes. Failing to accomplish this may result in the dissolution of the changes achieved through diligent work, with people reverting to their established and comfortable routines.

This approach to change has a deep foundation but requires strong leadership commitment and effort for the employees. If the changes are minor, Kotter's model may be too extensive. It is also important not to only perceive gamification as a general change process but also as a concrete implementation of software. To do this, it is common to use the so-called Quality Implementation Framework (QIF), which consists of four phases (Meyers, et al., 2012).

In the first phase of QIF, the clarification process takes place, where it is considered what needs and requirements the organization has. In these considerations, resources, adaptation possibilities, and capacity-building strategies are examined. With the involvement of stakeholders, a supportive environment is created to form an organization that can handle changes, including training employees. The second phase is about creating implementation teams and developing an action plan for implementation. Phase three is about having a supportive structure when the implementation starts. This can be technical assistance and support, coaching and supervision, as well as process guidance and supporting feedback mechanisms. Finally, in the fourth phase, work is done on improving future application and learning from experience. It is also here that the implementation should lead to standard practice.

Another approach has been a four-step REFA model with 1) preparation, 2) demonstration, 3) execution, and 4) completion developed by REFA. It has been suggested that there should be added an extra step of gamification between demonstration and execution (Kampker, et al., 2014).

Based on the ADKAR model, Kotter's 8-step model, the QIF model, and the enhanced REFA model, a 6-step model has been developed for implementing gamification at the workplace (see Figure 4.1).

1. Preparation and awareness. Here it is crucial to assess readiness, that is, to evaluate the organization's readiness for gamification

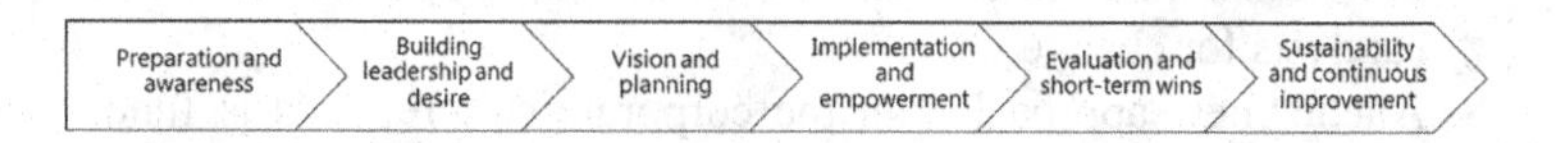

Figure 4.1 Six steps to implement gamification.

and identify barriers and obstacles. It is also at this step that all involved stakeholders are convinced to understand the benefits of gamification to achieve business goals.

2. Building leadership and desire. It is vital to ensure commitment and active involvement from senior leadership to champion the gamification initiative. This leads to forming a coalition with the necessary skills and influence to lead the gamification efforts, with the aim of motivating employees by highlighting the benefits of gamification such as increased engagement and productivity.
3. Vision and planning. Here a clear vision and strategy should be developed, defining the goals for gamification and how they align with organizational goals. This means developing an action plan with detailed implementation of gamification including timelines and responsibilities. It is also important to provide employees with information and training on how the gamification application will work so that all involved employees understand how to use it effectively.
4. Implementation and empowerment. This is where the action plan is being executed according to defined steps and timelines as well as identifying and eliminating barriers and obstacles that may hinder the implementation of gamification. At this step, it is crucial to encourage creativity and empower the employees to engage with the gamification solution, trying out the interface and actions built into the gamification application.
5. Evaluation and short-term wins. Get to achieve visible and understandable short-term successes with gamification with the aim of building momentum and demonstrating the effectiveness of the gamification application. Collect data on engagement, performance, and other relevant metrics to evaluate the impact of gamification. Also, celebrate and reward individuals and teams who actively participate and excel in gamification activities.
6. Sustainability and continuous improvement. This is the stage where gamification is integrated into everyday life and daily

routines, becoming part of the organizational culture, meaning that gamification should be anchored in the organization. In this connection, the new habits should be reinforced by introducing mechanisms that maintain gamification. However, it is also important that the gamification application is renewed and upgraded whenever there are changes in work tasks and routines so that the gamification application is continuously maintained, and there should also be ongoing consideration of opportunities to expand the benefits of gamification.

With these six steps it will be possible to do change management by implementing gamification into the organization in a way that leads to success. One could easily imagine that gamification would be so engaging and captivating that it would simply be popular just by throwing it into the organization. This, however, is a naïve stance, and just like any other organizational change and implementation of software gamification, implementation is not to be taken lightly. Only by a strategic effort to get people involved and change the organizational culture will gamification become a success if and only if the gamification application works and gives added benefits to the individuals and the organization.

Gamification design

The idea creation phase is about getting to understand the smart factory. The game designer will have to observe and interview workers at the smart factory to grasp the problems, in addition to finding out which problems can be solved by gamification. When the game designer has a mental map of some issues that need to be solved, then makes a model (it can be easily done with LEGO) of how the factory works and presents this model to the workers and discusses if the model is correct and try to come up with solutions. This is here an experienced game designer should use all his knowledge about different types of games and game mechanics to help come up with gamification solutions. When this is done, the first prototype is created, and of course there may be created several prototypes during the design process (McElroy, 2017; Fullerton, 2018).

Once a game designer is tasked with implementing gamification in an industrial setting, it is useful to draw on Greg Costikyan's understanding of games (see Chapter 1), following the idea that 'a game is an interactive structure that requires players to struggle toward goals' (Costikyan, 2002, p. 21).

Games have interactivity. This means that the player must have continuous interaction with the game. This can be problematic if the work processes are not interactive enough or cannot be properly registered. Games also have a structure, and so do work tasks. It is useful for the game designer to analyze the structure. For example, it could be repetitive manufacturing processes that resemble core gameplay loops (Marache-Francisco & Brangier, 2015).

Regarding struggle, it is crucial for gamification that there is a struggle taking place. This means that if a process is 99% correct, there is not much game in it. There must be choices and the possibility of failure for it to be an exciting game. Since the manufacturing industry strives for a zero-error culture (Kasulke & Bensch, 2017), gamification does not always work within that framework. One way to circumvent this is to create simulations or digital twins, where it is possible to fail without affecting production, creating a safe space for mistakes (Evans, 2017). Conversely, if it is possible to find work processes where struggle occurs, they can be well-suited for gamification.

Goals and objectives for games and industry may seem very similar but there are differences. The goal of a given manufacturing industry is to produce the necessary quantity of products for the market, and to achieve this, it goes through a long series of sub-processes in the industrial process until there is a finished product. The goal for the worker is to perform their tasks satisfactorily to achieve recognition, appreciation, and, of course, salary (Sirota & Klein, 2014). These goals must align with gamification, where the aim is to win by overcoming challenges along the way. As one can see, there are common threads, but the important thing in gamification is that the goal is clear and explicit, and it is exciting to reach the goal, which does not always match the manufacturing industry's objectives, where it is not about how exciting it is to reach the goal but merely fulfilling the goal and objectives. So, it requires a skilled game designer to align the industry's goals with the game's goals, where it becomes fun and where there is a higher degree of opportunity for learning while at the same time not only fulfilling but optimizing the manufacturing objectives.

Finally, we have Greg Costikyan's concept of endogenous meaning, where the game creates its internal logic. This can be both an advantage and a disadvantage because a game creates its frame where activities make sense and have meaning within that frame, but it does not necessarily have the same meaning outside that frame. In this way, one can ask if creating a game where completely different rules apply than in the real world. This phenomenon can add an extra layer of meaning to the experience when a work process has been gamified. The disadvantage can be that there arises a poor match between the real world and the game's logic. This can lead to the transfer of knowledge from the game to the real world becoming difficult because they are too far apart from each other and in the worst case means that the game's logic takes over in such a way that it becomes counterproductive, that is, that the game instead of optimizing processes creates counterproductive incentives. Hence, a game designer must align the game's endogenous meaning with the meaning inherent in the manufacturing industry, ensuring that the two conceptual frameworks converge and interact constructively.

Gamified work processes

As we have already established, the design of gamification for industry encompasses five key areas within production and logistics, characterized as follows (Warmelink, et al., 2020):

1. Product and process engineering.
2. Production planning.
3. Production execution and control.
4. Supply chain design and planning.
5. Transportation planning.

These five key areas can all undergo gamification design (see Chapter 3). The gamification design occurs through three approaches, namely affordances, psychological outcomes, and behavioral outcomes (Warmelink, et al., 2020). An understanding of all three components in the game design process is requisite for a game designer.

Affordances involve setting goals and objectives, as well as feedback based on metaphorical or fictive representations taking place so that they can drive the psychological outcomes in the right direction. This means that each time a design-related decision about affordances is made, the psychological impact and effect it has must be considered. Finally, it must lead to behavioral outcomes. Naturally, it is decisive that the affordances that create psychological effects ultimately lead to the consequence of promoting the desired behavior in industrial production.

In an analysis of a work process, a functional approach is to examine how the process begins and ends, and which patterns are repeated. This can be represented in a flowchart. When developers obtain an overview of the work procedure in this manner, it can be used to identify where typical errors and problems arise. That said, Ole Goethe points out there is a limit to what can be done, saying: 'Gamified thinking is not a miracle cure for all your product's woes. It's not possible to take a lousy process and sprinkle on some game elements and make it fantastic' (Goethe, 2019, p. 24).

The next step is to understand the work process as a game procedure, where game rounds are repeated, and significant choices are made. By viewing work processes as meaningful choices where the correct behavior is rewarded, it becomes possible to determine which game mechanics can be used to promote the correct behavior, motivate users, and foster learning – depending on the specific purpose of the gamification application.

In relation to flow, where the worker experiences a match between challenges and skills, it is important to design a gamification system where challenged employees in the beginning can easily obtain the difficulty level (Korn, 2012; Narayanan, 2014). This means ensuring that experienced skilled employees can make the learning curve less steep. Conversely, it applies so they experience flow. This can be done by creating more challenges and thus becoming able to learn in the workplace. It is important to create well-functioning games because poorly designed games or those that do not closely align with the specific needs and challenges of the workplace may fail to achieve their intended educational outcomes.

Another problem is resistance from employees who may be skeptical about such games, disbelieving the value of instructional games, either viewing them as frolicsome or simply inapt to their job functions. This is why it is important to design games that are obviously useful and come out as evidently relevant.

Our experience showed that about 5% of the workers in the factory were dedicated gamers who were so competitive that they were out of range compared to everyone else. This number might vary depending on the company culture, but the point is that designers should not expect everyone to be eager to play for the sake of the game itself.

It can be a trap for a game designer to assume that everyone is as passionate about playing as they are. This bias can obstruct the understanding of their audience. Consequently, employing Richard Bartle's categories of killers, achievers, socializers, and explorers presents certain challenges (Kocadere & Çağlar, 2018).

Firstly, Bartle's categories were originally developed for Multi-User Dungeons (MUDs) in the 1980s, where players shared a textual universe online. While this framework has since been applied with some success to understand players in massive multiplayer online role-playing games (MMORPGs), a workplace is not an online fantasy adventure. Secondly, not all employees are avid players. While it can be noted that most people play or are interested in playing, there is a difference between dedicated gamers and others who merely enjoy watching football and occasionally playing a board game or casual games on their smartphones. For this reason, the model has been improved into the so-called hexad player types consisting of philanthropists, disruptors, socializers, free spirits, achievers, and players (Tondello, et al., 2016; Lopez & Tucker, 2019).

Repetitive processes and core gameplay loops

In the context of shop floor gamification, it is crucial to manage repetitive processes effectively (Cherry, 2012). This approach is used in high-volume production because it ensures a consistent flow in the manufacturing process. For it to function properly, a high degree of standardization is required. Many industrial companies, such as car manufacturers, electronics factories, food and beverage production, and textile production, utilize repetitive processes. It is important to understand which loops are repeated, how they can be efficient and how consistent and predictable they are.

In game design, there are core gameplay loops (Martin, 2023). These often follow this model: The player performs an action, which can then be rewarded or possibly punished, leading to progression in the game. The player experiences this progression through feedback, which can then result in new challenges. Core gameplay loops repeat until the game is completed. The goal of the loop is to engage players, so they continue playing. This happens through flow, where the players' skills and challenges roughly match, ensuring that there is always a new challenge within the zone of proximal development (ZPD) (Hussain & Coleman, 2015; Gilbert, 2016).

Repetitive processes in industrial production are not the same as core gameplay loops, but they share some common traits and obvious similarities. First, both involve repeating activities or sequences and have a structured approach to how specific tasks and actions are performed in a certain order. Additionally, both make use of optimization and improvement, whether it concerns engagement or efficiency. In the manufacturing industry, the goal is to improve production speed, whereas in the gaming industry, the aim is to keep the player engaged with a series of interesting choices. Both approaches use feedback for this improvement.

However, the similarities end there. In the manufacturing industry, the goal of using feedback is, for example, quality control, with the aim of achieving uniformity and often high production speed (Mitra, 2016). Repetitive processes can also be highly automated and require minimal human activity, whereas core gameplay loops are about creating a gaming experience and, therefore, require human interaction. Consequently, the output is also very different. In games, the output of core gameplay loops is that the player or players experience progression in the game. In contrast, the output for manufacturing companies is naturally physical products or components.

As a game designer, the challenge is to identify repetitive processes that involve people and then pinpoint weaknesses in these processes where errors occur, or tasks and responsibilities are not performed optimally in relation to production efficiency. Once this is done, the goal is to determine how core gameplay loops can help enhance workers' job satisfaction and make them act more efficiently. This is achieved by adding game elements so that repetitive processes more closely resemble core gameplay loops.

The easiest way to add game elements is, of course, BLAP, which gives users feedback through points, badges, and achievements, and leaderboards often based on key performance indicators (KPI) (Heilbrunn, et al., 2014; Palenčárová, et al., 2022). When a task is performed suboptimal, no points are awarded, or there may even be a penalty. If the task is performed optimally or nearly optimally, points are awarded. However, if the goal is to create deep gamification, it may involve changing the whole structure so that the work feels more natural and enjoyable to perform.

Deep gamification can be achieved by adding a narrative structure, providing intricate challenges where the user experiences autonomous choices, and where each user can see personal development and progression (Dah, et al., 2024). As part of deep gamification, it may be necessary to include collaborative projects and healthy competition that create social interaction, and where feedback is not just simple rewards but rather constructive feedback, including bonuses, recognition, and career opportunities (Morschheuser, et al., 2017). All of this should support the purpose of deep gamification of reinforcing and developing the three underlying psychological requirements of self-determination in motivational psychology which are autonomy, competence, and relatedness.

Autonomy gives the worker control of one's own behavior and the freedom to choose, guided by self-initiation, self-interest, and personal values. Competence means the worker feels effective and capable of one's activities, gaining skill development, challenges, and feedback, and a sense of being able to achieve the desired outcomes and have an impact. And relatedness is about establishing meaningful relationships and interactions as well as a feeling of belonging while receiving and providing emotional support and understanding (Deci & Ryan, 2013).

This means that in a deeply gamified solution, everyday life becomes part of an overarching narrative, in which daily tasks and chores are included. Gameplay then becomes the execution of these tasks – preferably in collaboration with colleagues – and the goal is to gather knowledge about how the day went and analyze it, so that workers can receive constructive feedback on their performance and see how it went in relation to the overall mission and their own personal progression. Finally, the day can end with planning the goals for the next day together, with the intention of workers feeling a personal responsibility for the project.

With BLAP, the developers get an effective method for behavioral reinforcement, and it has an impact (Hamari, 2017; Burke, 2014). Nevertheless, BLAP may be combined with deep gamification in which meaningful tasks are provided, and the focus is not only on KPIs and competition but also on personal development. By combining these two approaches into a balanced gamification approach, the developer can create a gamification solution that employs both extrinsic and intrinsic motivation.

Aesthetics, customization, and personal profile

When a development team develops gamification, it is important that the aesthetic design is understandable and appealing to the target group. For example, if it is a simulation of some work processes, the simulation should be as close to reality as possible, both in terms of appearance and functionality. Game elements create joy, but it can be advantageous to present a professional or conservative aesthetic rather than wild colors, intense animations, and loud sounds (Dale, 2014). The reason for this design choice is that the game takes place in a serious context, and the game should be taken seriously, but also because the game should not attract undue attention as a game but instead support the already existing work processes. Nevertheless, if the gamification application has some deep gamification in the form of storytelling and narratives, it is also imperative to create aesthetics that fosters exploration, curiosity, and self-directed behaviors (Seo, et al., 2020; Jarrah, et al., 2024).

To create an easily understandable interface for users, it can be advantageous to use already well-known symbols (Caivano, 1998; Nitsche, 2008; Jappy, 2013; Jørgensen, 2013). Thus, icons such as bronze, silver, and gold are already established as familiar concepts for reward and success, and similarly, the colors of a traffic light – red, yellow, and green – are also well-established symbols to indicate legal or illegal actions. By using these established codes, it becomes much easier to explain the game's logic to people who are not necessarily hardcore gamers.

Customization is an opportunity for the workers to personalize the settings by choosing from different design options (Hicks, 2003; Nitsche, 2008; Tsourma, et al., 2019). The development team should create options that cater to design themes the workers can relate to. It could be general themes like sports, nature, and arts. This is also an opportunity for adding a picture of themselves and designing other kinds of virtual representations of themselves. Customization has the advantage of giving the users some kind of autonomy and empowerment.

Employees should have a personal profile where they can see their progress (Tsourma, et al., 2019). There should be an option to see how far they have come in various idle time games and how their scores are doing in learning while working gamification. Additionally, there should be a list of points, leaderboards, badges, and badge achievements, goal tracking, and other meaningful communication to the respective employee presented in an intuitive format. There may also be an option for customizing the personal profile to individual needs.

What is important to consider is whether one shares their profile with others. There are pros and cons to this. If developers make it very competitive, it should fit a competitive company culture, otherwise, it can do more harm than good, because while it may please the most skilled, it risks directly demotivating the less skilled, and worse, it risks destroying team cohesion and spirit. One way is that employees can only see their own progress and then decide for themselves if they want to share the information. In any case, it is important that the designer considers the implications it has and how it affects the company culture.

Instructional work-related games (IWRG)

Instructional work-related games are about creating games that can be played out in a safe space where the workers will learn. The process of designing such a game resembles that of any game, albeit with certain differences. The goal of the game is not ultimately to have fun and excitement. It is not a game where the game is an end to itself. It should not be understood that the players of IWRG cannot have fun during the process. While having a good experience is clearly advantageous, the primary focus is on the things they need to learn through the game.

As a game designer, it is, therefore, essential to conduct an analysis of the problems or issues the game should address. Once this is achieved, the game designer can proceed to develop a functioning game that can address and correct the problems. In such a scenario, it can be advantageous to use Costikyan's five design concepts: interactivity, structure, struggle, goals and objectives, and endogenous meaning. Beyond this, the game designer should also reflect on the aesthetic storytelling and game world that the game should adopt.

Idle time games (ITG)

Many of the same considerations apply when a game designer develops idle time games. Like with IWRG, the designer should understand the problems at the workplace that need to be solved. Game mechanics should be used to promote a psychological effect and, ultimately, desired behavior.

The difference is that the game takes place at the actual workplace, and this poses some requirements. First, there must be an opportunity to play ITG in peace and quiet. This means that time should be allocated for playing. This can possibly be done on a mobile phone or tablet, but this entails a security risk. Alternatively, it can be a computer that is provided, but in that case, it should not interfere with other processes at the workplace or obstruct those who want to play simultaneously.

The games should also be designed in such a way that it is quick to start and possible to stop at any time if assistance is needed or if the worker's skills are required.

This does not mean that they necessarily have to be simple casual games. That is, of course, an option, but it means that the games should be easy to jump into and out of when the opportunity arises.

Learning while working (LWW)

There are some prerequisites that need to be in place before developers can design LWW. The factory should have the relevant work processes digitized to such an extent that real-time gamification of tasks becomes possible. Any paperwork would enter the system with a delay, and feedback would be correspondingly delayed. Therefore, there will not be immediate feedback. Additionally, designers need access to sensitive production data, and it requires trust to gain such access.

Designers might consider supplementing with technology to support gamification, such as tablets or smartwatches. However, this requires that such additions are feasible. There may be parts of the production where other technology cannot be brought into the area due to safety reasons related to the production, but it could also be safety reasons concerning how data is stored and where the tablets, smartwatches, and similar devices are located when not in use. Therefore, such technological additions would need to be agreed upon in advance with the factory, both in relation to physical safety and data security.

Game engine

Instead of having to develop an entirely new gamification application from scratch each time, it can be advantageous to use a game engine (Ali & Usman, 2016; Lithoxoidou, et al., 2018). To begin with, developers can make use of game engines developed for the computer game industry, but they are optimized for producing impressive sound and graphics and a game mechanic primarily focused on a flow experience in the game. Many of the same basic elements can be used in gamification.

However, there may also be a reason to develop a game engine, sometimes called a gamification engine, that is optimized for gamification and serious games in the industry (Lithoxoidou, et al., 2018). In such a case, the game engine should have some of the most commonly used game mechanics built-in such as badges, leaderboards, levels, achievements, and points (see Chapter 1). The sounds and graphics should be optimized for the aesthetic that is appropriate in an industrial environment. Most importantly, it should be optimized to retrieve information from industrial control systems (ICS), such as SCADA. In this way, a lot of extra work can be avoided. Hence, the construction of a gamification engine for industrial environments should be targeted at validating its integration with the industry's main systems and auxiliary systems.

Additionally, such a game engine should be continuously updated with industrial standards, and it should be able to measure the performance of each individual employee, with inbuilt options for creating customizable profiles. Integration with existing data systems should be seamless, ensuring that real-time data can be used to adapt the game environment dynamically. Distinct requirements and ambitions inherent in various industries and organizations call for the gamification engine to show a high degree of adaptability to satisfy these specific needs. This necessitates providing malleability in the architecture of game mechanics, thereby permitting conveyed experiences that are consonant with corporate aims and employee functions. Multiple platforms, such as mobile devices, desktop computers, and wearable technology, ought to be supported by the engine to ensure accessibility and engagement throughout diverse work environments. By focusing on these valuable parameters, the games produced by the game engine optimized for gamification and serious games may cultivate an engaging and productive workplace, aiming at enhancing performance outcomes and increasing employee satisfaction.

Counterproductive incentives in gamification design

Counterproductive incentives or perverse incentives can occur when the full consequences of an incentive system are not thoroughly considered (Anguelov, 2015; Callan, et al., 2015). It is important to understand that as soon as a system is implemented to achieve a specific behavior, people will respond by seeking the system's benefits, but not necessarily by promoting the system's intention – unless they see a direct benefit in promoting the intended outcomes of the system. This refers to the embedded intention within the system. An incentive system can also have systemic flaws, meaning that even if everyone acts as intended, undesirable side effects can still arise (Gross & van de Lemput, 2021).

Developers can avoid counterproductive incentives by stress-testing game mechanics in their most extreme forms. This helps to identify systemic flaws. However, to address unwanted human behavior, it is necessary to understand human psychology and to test the system with people until the unwanted consequences of human behavior are discovered. Regrettably, this technique does not unfailingly catch every potential form of system corruption, thereby necessitating ceaseless monitoring for anomalies within the system. Regular oversight facilitates the modification of guidelines and the elimination of any loopholes that might lead to negative incentives.

A strategy to eliminate counterproductive incentives is to begin identifying potential counterproductive incentives. This involves stress testing the systems and gathering feedback from users in collaboration with behavioral psychologists and gamification specialists. Keep an eye out for signs of unintended behavior or demotivation. Also, ensure that the gamification aligns with the organization's values. Some counterproductive effects can be difficult to spot, so continuous monitoring of the gamification solution may be necessary. A rule of thumb is that if certain rewards are triggered suspiciously quickly and repeatedly, it might indicate that some users are gaming the system. Evaluate both quantitative and qualitative metrics. If there is only superficial engagement with low-quality work, this should be a red flag. Similarly, developers can assess the number of completed tasks and determine whether these metrics align with the quality of the work performed.

Cybersecurity and gamification in industrial production

When we talk about cybersecurity in factories, it is important to understand that these are high-risk companies that are continuously exposed to cyberattacks. In a time of uncertainty and war and an ongoing cyber war, where major powers like Russia and China plan and carry out cyberattacks on Western companies and institutions, it is crucial to understand that one cannot take cybersecurity lightly (Buchanan, 2020). Therefore, various defense techniques have been developed.

When it comes to security for the use of gamification systems in production, the Purdue Enterprise Reference Architecture (PERA) is commonly used (see Chapter 2). The PERA model has been groundbreaking for focusing on cybersecurity in companies (Mo & Beckett, 2018). Production environment integrity is maintained by the model, offering a systematic approach to the incorporation of cybersecurity measures within manufacturing enterprises. It confirms its effectiveness in the implementation of relevant security protocols, contributing to the defense of sensitive information and ensuring continuous operational activity.

At the lowest level, Level 0, there are no precautions in the model. However, companies can have their own rules. For gamification, it may become necessary to be on the shop floor to see how employees physically handle machines and understand how the machines work. It may also require observation, photography, and audio and video recordings for use in gamification. All of this poses a security risk and must be agreed upon with the company as to what is permissible.

At Level 1, we receive information from sensors about how production takes place in practice. Here, details are provided about all conceivable measurements that are necessary to monitor production. It will also show if there are breakdowns. This information is necessary for gamification on the shop floor, but it is also extremely sensitive. It is important to agree on what information is necessary, how it will be used, and how the gamification developers will handle security.

At Levels 3 and 4, it is about planning either in the form of manufacturing execution systems, batch management systems at Level 3, or enterprise resource management at Level 4. These are, of course, also sensitive and require access if gamification is to be conducted at these levels, and agreements need to be in place.

Regardless of the level at which information is necessary to implement gamification, it requires trust. It is important that the developers of gamification are very clear about what information they need for their gamification application. There must be a clear plan on how data is handled, who has access to the data, and what security rules need to be followed.

In any case, gamification will often require access to the deepest levels of the factory's security systems, as it will necessitate knowledge from either the direct processes or the historian database where the work processes are logged. Even if we implement the simple BLAP solution, this knowledge is necessary.

Ethical gamification design

When gamification is introduced to industrial work, it is not unproblematic and it raises some ethical questions (Sicart, 2009; Cherry, 2012; Raftopoulos, 2014). Accordingly, there ought to be ethical considerations concerning how gamification should be carried out. At the ethical level in persuasive design, one distinguishes between persuasion and coercion (Williams, 2024). Persuasion is when a system tries to promote positive behavior, whereas coercion is about getting people to perform actions or behaviors that are not in their interest or are otherwise unhealthy. The issue here is that the same psychological mechanisms based on behavior can be used for both persuasion and coercion. Therefore, the intention behind the system or any unintended side effects of the system, such as perverse incentives, must be considered, as these can lead to unethical use of gamification.

Designers working within gamification must be scrupulous in their efforts to balance performance focus, as overemphasis can lead to stress, burnout, and pervasive dissatisfaction among the users (Dolly, et al., 2024). Allowing competitive elements to become unduly dominant results in this. This is also seen when gamification is used to control employees to such an extent that there arises a division between employees. In coercive use of gamification of this sort, employees risk experiencing favoritism of groups or individuals due to performance rankings.

If there is too much focus on performance, the company risks undermining safety standards, so it should be considered in the design aimed at LLW gamification. It should not be such that employees take unnecessary risks to achieve rewards (Şenol & Onay, 2023). In addition to this, there must also be a focus on ensuring that gamification does not lead to a form of gambling addiction (Andrade, et al., 2016). It should also not be the case that gamification in the long term destroys intrinsic motivation because the whole gamification system is based on extrinsic motivation (Chou, 2019). In the worst case, this can lead to counterproductive incentives to only perform functions if there is a reward, whereas previously they helped each other without a reward because there was intrinsic motivation to be a good colleague. Another ethical issue is privacy. It concerns how the collection and monitoring of data takes place, and it is a careful handling of data. It must be ensured that data is not misused (Mavroeidi, et al., 2019). Furthermore, the data ought to be secured so that it remains inaccessible to individuals without proper authorization (Manjikian, 2023).

It would be ethically appropriate for gamification solutions to endorse an ethical framework. This means that systems not only promote engagement in performing tasks but also correct ethical behavior according to work rules and social behavior with colleagues. The game designer should meticulously integrate these ethical principles into the gamification design, ensuring that users are encouraged to act in a morally responsible manner. The game designer must consistently evaluate and update this ethical alignment to stay attuned to the progression of workplace standards and societal norms.

Summary

Creating gamification and serious games in Industry 4.0 and smart manufacturing can be done in the forms of IWRG, ITG, and LWW. Gamification is a variant of experience design with a particular focus on incorporating game mechanics from computer games. To implement gamification, it is necessary to have a team, where the leadership consists of a project manager, a game designer, and preferably also a programmer and an engineer. Depending on the project's size, additional engineers, programmers, and designers may be needed.

In developing gamification for manufacturing companies, it is crucial to communicate and understand what happens in the workplace, where it is possible to make improvements in terms of motivation and productivity. This requires a change management strategy to successfully implement gamification. Furthermore, developers must analyze and understand which work processes gamification can positively impact and develop games in a style that fits the company culture, where each employee can track their progress in a personal profile. It may be advantageous to develop a dedicated game engine for gamification.

Finally, the development team must ensure that their gamification solution does not create undesirable counterproductive incentives and ensure that data remains secure. All of this should be evaluated with an eye toward ensuring that gamification solutions are ethically responsible.

Bibliography

Achterbergh, J. & Vriens, D., 2019. *Organizational Development: Designing Episodic Interventions.* New York, NY: Routledge.

Ali, Z. & Usman, M., 2016. *A Framework for Game Engine Selection for Gamification and Serious Games.* San Francisco, CA: IEEE. https://ieeexplore.ieee.org/stamp/stamp.jsp?arnumber=7821753

Andrade, F. R. H., Mizoguchi, R. & Isotani, S., 2016. *The Bright and Dark Sides of Gamification.* Cham, CH: Springer.

Anguelov, N., 2015. *The Dirty Side of the Garment Industry: Fast Fashion and Its Negative Impact on Environment and Society.* Boca Raton, FL: CRC Press.

Applebaum, S. H., Habashy, S., Malo, J.-L. & Shafiq, H., 2012. Back to the future: revisiting Kotter's 1996 change model. *Journal of Management Development*, 31(8), pp. 764–782.

Badiru, A. B., 2019. *Systems Engineering Models: Theory, Methods, and Applications.* Boca Raton, FL: CRC Press.

Buchanan, B., 2020. *The Hacker and the State: Cyber Attacks and the New Normal of Geopolitics.* Cambridge, MA: Harvard University Press.

Burke, B., 2014. *Gamify: How Gamification Motivates People to Do Extraordinary Things.* New York, NY: Routledge.

Cadle, J. & Yeates, D., 2008. *Project Management for Information Systems.* Essex, UK: Pearson Education.

Caivano, J. L., 1998. Color and semiotics: A two-way street. *Color Research & Application*, 23(6), pp. 390–401.

Callan, R. C., Bauer, K. N. & Landers, R. N., 2015. How to Avoid the Dark Side of Gamification: Ten Business Scenarios and Their Unintended Consequences. In: *Gamification in Education and Business.* Cham, CH: Springer, pp. 553–568.

Cherry, M. A., 2012. The gamification of work. *Hofstra Law Review*, 40(4), pp. 851–858.

Chou, Y.-k., 2019. *Actionable Gamification: Beyond Points, Badges, and Leaderboards.* Birmingham, UK: Packt Publishing.

Clark, D., 2021. *Learning Experience Design: How to Create Effective Learning that Works.* London, UK: Kogan Page Publishers.

Cockburn, A., 2007. *Agile Software Development: The Cooperative Game.* Boston, MA: Pearson Education.

Costikyan, G., 2002. *I Have No Words & I Must Design: Toward a Critical Vocabulary of Games.* Tampere, FI: Tampere University Press, pp. 9–33.

Dah, J., Hussin, N. & Aliu, A. A., 2024. Gamification Is Not Working: Why?. In: Games and Culture. Thousand Oaks, CA: SAGE Publications.

Dale, S., 2014. Gamification: Making work fun, or making fun of work?. *Business Information Review*, 31(2), pp. 82–90.

Deci, E. L. & Ryan, R. M., 2013. *Intrinsic Motivation and Self-Determination in Human Behavior*. New York, NY: Springer Science-Business Media.

Dollins, M. & Stemmle, J., 2021. *Engaging Employees through Strategic Communication: Skills, Strategies, and Tactics*. New York, NY: Routledge.

Dolly, M., Nimbarte, A. & Wuest, T., 2024. The effects of gamification for manufacturing (GfM) on workers and production in industrial assembly. *Robotics and Computer-Integrated Manufacturing*, 88, pp. 1–20.

Douglass, B. P., 2021. *Agile Model-Based Systems Engineering Cookbook: Improve System Development by Applying Proven Recipes for Effective Agile Systems Engineering*. Birmingham, UK: Packt Publishing.

Evans, M., 2017. Providing students with real experience while maintaining a safe place to make mistakes. *Association for Journalism Education*, 6(1), pp. 76–83.

Fullerton, T., 2018. *Game Design Workshop: A Playcentric Approach to Creating Innovative Games, Fourth Edition*. Boca Raton, FL: CRC Press.

Gilbert, S., 2016. *Designing Gamified Systems: Meaningful Play in Interactive Entertainment, Marketing and Education*. Burlington, MA: Focal Press.

Goethe, O., 2019. *Gamification Mindset*. Cham, CH: Springer.

Gomaa, H., 2011. *Software Modeling and Design: UML, Use Cases, Patterns, and Software Architectures*. Cambridge, MA: Cambridge University Press.

Gross, L. & van de Lemput, C., 2021. Threats Arising from Software Gamification. In: *The Role of Gamification in Software Development Lifecycle*. London, UK: BoD – Books on Demand, pp. 21–46.

Hamari, J., 2017. Do badges increase user activity? A field experiment on the effects of gamification. *Computers in Human Behavior*, 71, pp. 469–478.

Heilbrunn, B., Herzig, P. & Schill, A., 2014. *Tools for Gamification Analytics: A Survey*. London, UK: IEEE/ACM.

Herzig, P., Ameling, M., Wolf, B. & Schill, A., 2015. Implementing Gamification: Requirements and Gamification Platforms. In: *Gamifi cation in Education and Business*. Cham, CH: Springer, pp. 431–450.

Hiatt, J. M. & Creasey, T. J., 2003. *Change Management: The People Side of Change*. Loveland, CO: Prosci.

Hicks, D., 2003. Supporting personalization and customization in a collaborative setting. *Computers in Industry*, 52(1), pp. 71–79.

Hilburn, T. B. & Towhidnejad, M., 2021. *Software Engineering Practice: A Case Study Approach*. Boca Raton, FL: CRC Press.

Hussain, T. S. & Coleman, S. L., 2015. *Design and Development of Training Games*. New York, NY: Cambridge University Press.

Jappy, T., 2013. *Introduction to Peircean Visual Semiotics.* London, UK: Bloomsbury Academic.

Jarrah, H. Y. et al., 2024. The impact of storytelling and narrative variables on skill acquisition in gamified learning. *International Journal of Data and Network Science*, 8, pp. 1161–1168.

Jørgensen, K., 2013. *Gameworld Interfaces.* Cambridge, MA: The MIT Press.

Kampker, A. et al., 2014. *Increasing Ramp-Up Performance By Implementing the Gamification Approach.* Amsterdam, NL: Elsevier.

Kasulke, S. & Bensch, J., 2017. *Zero Outage: Putting ICT Quality First in the Digital Era.* Cham, CH: Springer.

Keepers, M., Nesbit, I., Romero, D. & Wuest, T., 2022. Current state of research & outlook of gamification for manufacturing. *Journal of Manufacturing*, 64, pp. 303–315.

Kocadere, S. A. & Çağlar, Ş. Ç., 2018. Gamification from player type perspective: A case study. *Educational Technology & Society*, 21(3), pp. 12–22.

Korn, O., 2012. *Industrial Playgrounds: How Gamification Helps to Enrich Work for Elderly or Impaired Persons in Production.* New York, NY: ACM.

Kotter, J. P., 2012. *Leading Change.* Boston, MA: Harvard Business Review Press.

Leclercq, T. et al., 2020. When gamification backfires: the impact of perceived justice on online community contributions. *Journal of Marketing Management*, 36(5–6), pp. 550–577.

Lithoxoidou, E. E. et al., 2018. *A Gamification Engine Architecture for Enhancing Behavioral Change Support Systems.* Corfu, GR: ACM.

Liu, D., 2018. *Systems Engineering: Design Principles and Models.* Boca Raton, FL: CRC Press.

Lopez, C. E. & Tucker, C. S., 2019. The effects of player type on performance: A gamification case study. *Computers in Human Behavior*, 91, pp. 333–345.

Manjikian, M., 2023. *Cybersecurity Ethics: An Introduction.* Second ed. New York, NY: Routledge.

Marache-Francisco, C. & Brangier, É., 2015. Gamification and human-machine interaction: A synthesis. *Le Travail Humain*, 78, pp. 165–189.

Martin, J., 2023. *The Ludotronics Game Design Methodology: From First Ideas to Spectacular Pitches and Proposals.* Boca Raton, FL: CRC Press.

Mavroeidi, A.-G., Kitsiou, A., Kalloniatis, C. & Griyzalis, S., 2019. Gamification vs. privacy: Identifying and analysing the major concerns. *Future Internet*, 11(3), pp. 1–17.

McElroy, K., 2017. *Prototyping for Designers: Deeloping the Best Digital and Physical Products.* Sebastopol, CA: O'Reilly Media.

Merholz, P. & Skinner, K., 2016. *Org Design for Design Orgs: Building and Managing In-House Design Teams.* Sebastopol, CA: O'Reilly Media.

Meyer, B., 2014. *Agile!: The Good, the Hype and the Ugly.* Cham, CH: Springer.

Meyers, D. C., Durlak, J. A. & Wandersman, A., 2012. The quality implementation framework: A synthesis of critical steps in the implementation process. *American Journal of Community Psychology*, 50, pp. 462–480.

Mitra, A., 2016. *Fundamentals of Quality Control and Improvement.* Fourth ed. Hoboken, NJ: John Wiley & Sons.

Mo, J. & Beckett, R., 2018. *Engineering and Operations of System of Systems.* Boca Raton, FL: CRC Press.

Morschheuser, B., Maedche, A. & Walter, D., 2017. *Designing Cooperative Gamification: Conceptualization and Prototypical Implementation.* Portland, OR: ACM.

Narayanan, A., 2014. *Gamification for Employee Engagement.* Birmingham, UK: Packt Publishing.

Nitsche, M., 2008. *Video Game Spaces: Image, Play, and Structure in 3D Worlds.* Cambridge, MA: The MIT Press.

Palenčárová, J., Abuladze, L. & Blštáková, J., 2022. *Goal Setting and KPI Measurement as Tools for Broader Use of Online Gamification.* Bratislava, SK: University of Economics in Bratislava.

Pedersen, R., 2009. *Game Design Foundations.* Second ed. Sudbury, MA: Jones & Bartlett Publishers.

Raftopoulos, M., 2014. Towards gamification transparency: A conceptual framework for the development of responsible gamified enterprise systems. *Journal of Gaming & Virtual Worlds*, 6(2), pp. 159–178.

Reis, A. C., Silva Júnior, E., Gewehr, B. B. & Torres, M. H., 2020. Prospects for using gamification in Industry 4.0. *Production*, 30, e20190094. https://doi.org/10.1590/0103-6513.20190094.

Routledge, H., 2016. *Why Games Are Good For Business: How to Leverage the Power of Serious Games, Gamification and Simulations.* New York, NY: Palgrave Macmillan.

Salzman, H. & Rosenthal, S. R., 1994. *Software by Design: Shaping Technology and The Workplace.* Oxford, UK: Oxford University Press.

Savignac, E., 2017. *The Gamification of Work: The Use of Games in the Workplace.* London, UK: ISTE.

Schensul, S. L., Schensul, J. J. & LeCompte, M. D., 1999. *Essential Ethnographic Methods: Observations, Interviews, and Questionnaires.* Walnut Creek, CA: Altamira.

Şenol, D. & Onay, C., 2023. Impact of gamification on mitigating behavioral biases of investors. *Journal of Behavioral and Experimental Finance*, 37, pp. 1–12.

Seo, K. et al., 2020. Goldilocks conditions for workplace gamification: How narrative persuasion helps manufacturing workers create self-directed behaviors. *Human-Computer Interaction*, 36(5–6), pp. 473–510.

Sicart, M., 2009. *The Ethics of Computer Games.* Cambridge, MA: The MIT Press.

Sirota, D. & Klein, D. A., 2014. *The Enthusiastic Employee: How Companies Profit by Giving Workers What They Want.* Second ed. Upper Saddle River, NJ: Pearson.

Suhartini, R., Ramadhani, B. Y. & Wahyuningsih, U., 2024. Improving teaching factory performance by work culture in vocational learning. *Eurasian Journal of Educational Research*, 109, pp. 236–249.

Tennant, D. V., 2022. *Product Development: An Engineer's Guide to Business Considerations, Real-World Product Testing, and Launch.* Hoboken, NJ: John Wiley & Sons.

Thompson, D., 2018. *The Human Factor in Project Management.* New York, NY: Auerbach Publications.

Tondello, G. F. et al., 2016. *The Gamification User Types Hexad Scale.* New York, NY: ACM.

Tsourma, M. et al., 2019. Gamification concepts for leveraging knowledge sharing in Industry 4.0. *International Journal of Serious Games*, 6(2), pp. 75–87.

Tyler, J., 2015. *Building Great Software Engineering Teams: Recruiting, Hiring, and Managing Your Team from Startup to Success.* New York, NY: Apress.

Ulmer, J., Braun, S., Cheng, C.-T. & Wollert, J., 2022. Usage of digital twins for gamification applications in manufacturing. *Procedia CIRP*, 107, pp. 675–680.

Visser, J., Rigal, S., Wijnholds, G. & Lubsen, Z., 2016. *Building Software Teams: Ten Best Practices for Effective Software Development.* Sebastopol, CA: O'Reilly Media.

Warmelink, H., Koivisto, J., Mayer, I. & Vesa, M., 2020. Gamification of production and logistics operations: Status quo and future directions. *Journal of Business Research*, 106, pp. 331–340.

Williams, J., 2024. Ethical dimensions of persuasive technology. In: *Oxford Handbook of Digital Ethics.* Oxford, UK: Oxford University Press, pp. 281–291.

Conclusion

Gamification in Industry 4.0 integrates game elements into an industrial context with smart manufacturing. Normally, gamification has been introduced as shallow gamification based on extrinsic motivation in the form of BLAP. However, there has also been a wish to create deep gamification in which intrinsic deep motivation can enhance a fulfilling experience. Addressing both extrinsic and intrinsic motivation requires the implementation of a balanced gamification strategy that combines shallow and deep gamification, integrating both practices for a functional strategy aligned with motivational psychology and learning theory. The gamification for industrial work must be integrated into Industry 4.0 technologies such as AI, Big Data, Cloud Computing, and Digital Twin, all of which are key technologies for smart manufacturing.

There are different gamified strategies based on serious games and gamification. Instructional work-related games (IWRG) are based on serious games executed outside the parameters of industrial production. Idle time games are serious games or gamification that can be played during idle times for the workers at the factory, and finally, there is learning while working (LWW) in which gamification is implemented into the daily work with the possibility of real-time feedback.

For fruitful gamification, a dedicated team consisting of a project manager, game designer, programmers, and engineer is indispensable, as well as having a robust understanding of workplace dynamics and the execution of a change management strategy is crucial for achieving effective and successful implementation. The serious games and gamification should be aligned with company culture and track

DOI: 10.4324/9781003406822-5

employees' progress through personal profiles. Also, developing an exclusive gamification engine has the potential to produce promising gamification design solutions.

Ensuring the safeguarding of data and the prevention of counter-productive incentives must be prioritized as ethical considerations. Ramifications of gamification should be meticulously evaluated to ensure ethical responsibility and the achievement of positive outcomes devoid of negative unintended consequences. Adequately implemented gamification can significantly elevate motivation, productivity, and learning in intelligent manufacturing contexts.

To develop gamification and serious games that promote motivation, productivity, and learning in Industry 4.0 and smart manufacturing, it is essential to integrate both shallow and deep gamification strategies, align them with AI, Big Data, Cloud Computing, and Digital Twin technologies, and ensure they are tailored to the company culture and ethics while providing real-time feedback and tracking employee progress through personal profiles.

Index

Printed in the United States
by Baker & Taylor Publisher Services

Printed in the United States
by Baker & Taylor Publisher Services